THE IMPERILED OCEAN

THE IMPERILED OCEAN

HUMAN STORIES FROM A CHANGING SEA

LAURA TRETHEWEY

PEGASUS BOOKS
NEW YORK LONDON

THE IMPERILED OCEAN

Pegasus Books Ltd.
148 W 37th Street, 13th Floor
New York, NY 10018

First Pegasus Books edition November 2019

Interior design by Maria Fernandez

Library of Congress Cataloging-in-Publication Data is available.

ISBN: 978-1-64313-198-6

10 9 8 7 6 5 4 3 2 1

Printed in the United States of America
Distributed by W. W. Norton & Company

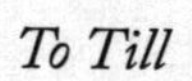

To Till

CONTENTS

INTRODUCTION

LEARNING TO FLOAT

We reached this place by water. Hours ago we'd left the last paved road and driven into dark rain forest, gravel boiling at our tires, thick evergreens flashing past, and stars coming out overhead. It was morning when we reached the end of the road. The forest fell away and the Pacific Ocean took center stage. The wind freshened a few knots. The temperature dropped a few degrees. The sun hit the water and made a shimmery band on the surface, following my eye wherever I moved, like the ocean was waiting just for me.

Down by the docks, we met a wide-set man, blind in one eye, who operated a water taxi service between the end of the fire road and his hometown, a half-hour boat ride away. We climbed on board, and the driver, who introduced himself as Leo, jetted us away to Kyuquot, an unincorporated Indigenous village clinging to the shores of a channel in the rugged Pacific Northwest. On starboard, forested mountains rose majestically; on port, a spectacular, drop-off-the-edge-of-the-Earth view of wide-open ocean that caught my breath every time. I breathed in air that smelled as clean and fresh as a chopped cucumber. Nothing but ocean from here to Japan.

This ecological reserve, the traditional territory of the Kyuquot and Checleseht First Nations, appears pristine, untouched, and eye-wateringly beautiful, especially on a sunny day. I had joined a volunteer team to collect plastic on MuQwin Peninsula, a mountainous cape that sticks straight out into the ocean. Because of MuQwin's geography and the year-round storms that slam in, the cape acts like a giant sieve for ocean plastic along the West Coast of North America. It has one of the most plastic-polluted shores in Canada, a country with more coastline than any other in the world.

We weren't even in MuQwin yet, but I could already see evidence of a plastic hot spot onshore. Nets, foam, bottles, fishing buoys, flip-flops, tennis balls—a mere scintilla of the eight million metric tons of plastic that leak from land to sea every year—strewed the coastline. This lonely place with a few scattered villages might feel distant, faraway, and untouched by the wider world, but the garbage washing in told a global story: sea-ravaged labels come in Japanese, Danish, Turkish, Taiwanese, and Greek.

It wasn't always this way. Leo remembered how the ocean once served as his town's dump. Most of his life Kyuquot had no money to pay for garbage collection, and the town's waste ran from land to sea, just as it still does in many developing countries around the world. Leo tried to attract media attention by focusing on the leatherback turtle, a giant of a reptile that migrates all the way from Indonesia to feed in the jellyfish-rich waters of the Pacific Northwest. Instead they feasted on plastic bags that float hypnotically through the water looking just like the turtles' prey. A few years ago, the town cleaned up the inlet when an outsider helped introduce a garbage pickup service. The water finally cleared, but now waste was flowing in from a whole new front: the open ocean.

This story about a village by the sea, a complicated past behind it, a challenging future ahead, is like so many stories I've heard about the ocean. The details shift, and so do the winners and losers, but the theme of unavoidable change is omnipresent, change so deep and wide-reaching that it is beyond the ken of any single person and raises a larger question for me: what is going on out at sea?

The poet Anne Stevenson once wrote, "The sea is as near as we come to another world." I feel the truth of these words when diving in the ocean, listening to the fragile in-and-out of my breath against the vast silence of the sea, watching the watery murkiness resolve into a new world so unlike my own. And when I read stories about copulating sea slugs with thorny penises, or camera-harnessed beluga whales suspected of Russian espionage, or of an offshore sailor in the Indian Ocean watching a fireball fall from the night sky, claiming to have witnessed the last moments of missing Malaysia Airlines flight MH370, I feel like I am poring over alien transmissions from another planet. Something huge and shadowy and important is going on out *there* in the lifeblood of this planet, the world's greatest ecosystem, the hidden force behind infrastructure, weather, and our daily small talk about the weather. The watery surface is a place of transit and trade; the seafloor a place of connection, finance, communication, and untold riches. All these unfathomable connections lead to a greater story of change just beyond the horizon.

Stories about the sea are different. The stakes are higher, the sacrifices dearer, and the rewards richer for those who risk crossing the water. The people who are out on the ocean every day know its life better than anyone else, people of all races, motivations, and professions who have the most up-to-date insight into the last great mystery here on Earth. In 2015, I set out on an extended listening tour to find out what people were doing in this vast space that we know so little about. I started with a simple question: what do people want from the ocean? I was drawn to refugees crossing the Mediterranean on barely seaworthy boats; to cruise ship workers disappearing overboard; to hobby sailors cruising off into storms; to a cinematographer shooting underwater scenes for blockbuster movies; to a water-dwelling community fighting eviction; and to a biologist tracking an ancient species that was being erased from the water. I was curious about how someone's stature affected success at sea; how larger constructs, like laws, nationality, storytelling, and science, worked on water. I had a suspicion that the unbelievable stories that drifted back to land ran much deeper, caused by a friction between natural and man-made rules. The only way to find out if I was right was to talk to people, to hear stories of what they were experiencing. Only then might I begin to piece together the larger story of a changing ocean. Each person I met along the journey gave me insight into a question that turned out to be anything but simple.

It would be impossible to catalogue in a single book all the seven and a half billion reasons humans are drawn to the sea. I split my focus into motivations: working, migrating, cleaning, creating, adventuring, and researching. The most common motivation for going to sea, unsurprisingly, was money. According to the National Ocean Service, every year three million Americans work on the ocean, in fishing, oil and gas, tourism, the Coast Guard, and many other industries. Multiply that number by a thousand and you get three billion people worldwide whose livelihoods, the United Nations Environment Programme tells us, depend on the ocean.

Another obstacle I encountered was that as soon as I said "ocean," people assumed I was writing a book that would appeal only to die-hard environmentalists. I found that extraordinary. So much more is going on at sea than what is tidily compartmentalized as environmental. Shipping routes wrap around the globe like a ball of twine, transporting over 90 percent of the objects we use every day. We pump oil—the modern world's lifeblood—from the seafloor, and in the future we plan to harvest even more energy through offshore wind farms and wave power. The majority of internet traffic, financial trades, and Google searches hum along fiber-optic cables laid across the seafloor. When I tried to interview an Alaskan fiber-optic cable company installing the first subsea cable through the legendary Northwest Passage, its public relations team probed for my angle. They asked if I was writing a story about the impact of sonar mapping and noise on marine mammals. Yes, I said, I was curious about that, but I wanted to know about the Arctic communities that would benefit from high-speed internet and the challenges, both financial and technical, of installing infrastructure in an ice-locked waterway that is only now, thanks to a changing climate, becoming navigable. The company connected me with a ship captain who was laying their fiberoptic cable along the seafloor. The captain couldn't figure me out, and later the company stopped returning my messages. Not long after, the company helped to report its CEO to authorities for defrauding investors of over $250 million in order to fund the first phase of the project—a charge to which she later pled guilty. Stories at sea *are* different. They're deeper and richer than what we imagine from land.

Four and a half billion years ago, the Earth was born with a liquid surface, molten lava rolling around the world in great unbroken waves. In the beginning, there were no landmasses to collide against and no atmosphere to rain down water. Eventually, as the surface of the Earth cooled and hardened, water arrived. And with it, came life.

Scientists continue to debate how 71 percent of the Earth became submerged in water, creating 99 percent of the habitable space available on the planet by volume. But the most commonly accepted theory for how life began proposes that billions of years ago, life emerged out of a crack in the seafloor, in the deepest, darkest, most inhospitable conditions on Earth. A few individual cells fed on chemicals produced when superhot slurry of volcanic salt water, heated by the Earth's magma core, met supercold salt water at the seafloor. This collision created a chemical reaction that generated energy for life to grow, just as sunlight makes plants flourish.

In the fossil record, it's clear that life in water came first: the oldest animals on earth are aquatic. Early primitive single-celled organisms needed to live in water, while today more complex multicellular organisms have evolved to carry water around inside us. We didn't know how water gave birth to life until 1977, when a group of geologists from the Woods Hole Oceanographic Institution in Massachusetts hit scientific pay dirt. The research team set out on a boat two hundred kilometers off the coast of Ecuador, looking for a hydrothermal vent that they speculated might emerge between the shifting plates of Earth's surface—the theory of plate tectonics still being new and revolutionary at the time When they went to plumb the depths with a tiny submarine, they found lobsters, crab, anemones, shrimp, mussels, and thousands of other organisms gathered around a hydrothermal vent. This rip in the Earth's seam spewing chemicals into the water created an underwater Garden of Eden—perhaps the secret to life's beginnings on Earth.

Maybe we've been terrestrial animals for too long. Our entire evolutionary history took place on land, so we tend to forget this watery origin story. In natural history museums, we favor a land-based portrayal of our survival with pictures of early humans tracking mastodons across wide-open savannahs. But, as the work of archaeologists Jon Erlandson and Scott Fitzpatrick tells us, *Homo erectus* and *Homo sapiens* flourished along African coastlines, foraging for shellfish and surviving off creatures left behind in the intertidal. These first modern humans may even have followed the coastline out of Africa as they expanded to Asia and beyond. In a cave on the South African coast, a group of spelunking archaeologists from the University of Arizona discovered the remains of a mussel, clam, and barnacle dinner dating back 164,000 years, when much of the world was still locked in glaciers. These seafood dinners, the archaeologists speculated, might have ensured our survival through inhospitable cold periods. Not only was the ocean our birthplace but it was our cradle as well.

Thousands of years later, the vast majority of us still rely on the ocean in some form or another. A third of the world's population lives within sixty miles (one hundred kilometers) of a coastline. Global fisheries unload 170 million metric tons of seafood at the dock every year, both farmed and wild caught, which translates into $143 billion annually in export value. Every day, three billion people consume seafood for as much as 20 percent of their animal protein intake. That number is on the rise: in 2016, humans ate more fish than ever before and our yearly consumption of seafood is growing faster than that of all land animals combined. More than 30 percent of the world's oil and gas is extracted offshore, while coastal tourism contributes 6–7 percent of worldwide employment. Taken together, the ocean's global economic activity is worth an estimated 3–6 trillion dollars a year, rivaling the gross domestic product of the world's biggest national economies.

When you don't live close enough to the sea to feel its damp chill or hear the call of a foghorn daily, the ocean can become a foreign concept, a collection of sensations rather than a real place. Inlanders might feel that coastal communities are the sole beneficiaries of living near water, with its seafood, trade, tourism, and even rising real estate prices. Of course, coastal communities have to contend with the trade-offs of destructive weather and rising sea levels, expensive rent, and even corrosive salty winds that eat away at bikes and cars. When we factor in the human-generated carbon emissions that the ocean takes up from the atmosphere and the heat that it tempers; the oils, minerals, and metals mined; our shared evolution, cultural history, and every second breath we take—half of the Earth's oxygen is produced by the ocean's bright green phytoplankton—it's clear we all have a connection to the sea no matter where we live.

When the first European sailing nations set out to explore the sea and whatever lay beyond the horizon, the open ocean was unmapped, unknown, assumed to be hostile. The Atlantic Ocean was known as *Mare Tenebrosum*: Latin for "Sea of Darkness," *Bahr al-Zulamat* in Arabic. Similar feelings of disdain or foreboding crop up in India, where Hindu law prohibits voyages by sea; in ancient Greece and Byzantium, where contempt for sailors and sea trade was commonplace; and in the 1500s, when the Portuguese dominated seafaring but very few countrymen actually wanted to sail far from land. In early human mythology, the ocean was often imagined as an angry or benevolent god, taking away lives with one hand and giving life with the other. Without accurate weather predictions, earlier coastal communities had no idea when the ocean might switch from tranquil to terrifying. Hurricanes and typhoons rose up without warning, washing away lives and livelihood. Seagoing Vikings and other marauders snuck up with little warning and did the same. Even today, with all the improvements to modern weather forecasting, a storm catching you unawares on the water is a fearsome experience.

Talking to diverse groups about going to sea, I saw a division between ocean crossings that were voluntary and involuntary. If we feel prepared for the journey and come to the sea with a sense of adventure, we feel connected to and energized by the ocean. "I think the best way to describe the experience would be as 'total immersion,'" one sailor put it after sailing twenty-five thousand nautical miles across the Pacific Ocean. "Twenty-four-seven bombardment of the senses. Not in a bad way—in fact, often in a fantastic way—but a bombardment nonetheless." On smaller vessels, the ocean demands full immersion: the smells, the sounds, the constant motion, as well as an adrenaline-fueled awareness that at any moment, the ocean could turn ugly. For those who make a crossing out of necessity or a misguided sense of romanticism, the ocean becomes a sort of prison, something to be endured until land heaves into sight.

From an early age, we hear stories about the ocean that ward us off or draw us in. Whatever we see in this vast space, those early stories take on an outsized role in our imagination, shaping a lifelong connection to water. The ocean is always laced with human fears and ambitions. Roman and Greek gods embodied the ocean and directed its powers for good or evil. Sirens led sailors to their death. Aquatic dinosaurs haunted Scottish lochs. Ancient underwater cities held untold treasures buried somewhere in the depths below. In keeping with the romantic spirit of her age, Charlotte Brontë fainted at her first brush with the ocean. J. M. W. Turner lashed himself to a ship's mast for hours to paint *Snow Storm*, one of the most expressive ocean images ever put onto canvas. Later in the 19th century, the Victorians came to see the water as a recreational space, a place for rejuvenation in resort towns built along the coastline. In the 1970s, *Jaws* scarred an entire generation.

In 1603, a Dutch captain attacked and seized the goods of a Portuguese galleon sailing in the East Indies. Both the Portuguese and the Spanish claimed sovereignty over the sea at the time, but a few years later, the Dutch jurist Hugo Grotius wrote a carefully reasoned defense of the Dutch captain that he called *Mare Liberum* (Freedom of the Seas). He argued that no nation or person could own the ocean, since it was a naturally and historically unoccupied space that was used by all. Thus, the ocean was free and should remain so for eternity. "Paradoxically, this defence of an act of piracy turned out to be the starting point for modern international maritime law," notes William Langewische in *The Outlaw Sea*.

This idea of freedom at sea persists. The United Nations Convention on the Law of the Sea carves the water into nationalistic little chunks determined by the nearest country's coastline. For centuries, most coastal countries followed the canon-shot rule that held that territorial seas extended three nautical miles from land (3.5 miles, or 5.5 kilometers), or about as far as a cannon ball could fly—which gives a sense of how long ago these rules were written. In the 1980s, most countries extended their jurisdiction to twelve nautical miles (13.5 miles, or 22 kilometers). In this space, the country is king, ruling over the sky above and the water below. How a country uses its territorial seas, as well as the outlying ocean jurisdictions, is often a signifier of national priorities. Tiny island nations like the Pacific island New Caledonia or the Caribbean islands of Antigua and Barbuda have established massive marine protected areas to protect their ocean resources. This makes sense in countries where water outstrips land. In Norway, where the salmon is revered, massive salmon farms operate offshore and millions of fish swim in digitally monitored pens. In the Arctic Ocean, Canadian hydrographers are confirming where exactly the seafloor drops off to depths beyond 2,500 meters, which could extend Canada's access to resources in the contested waters of the North Pole. The farther out to sea you go, the more the nearest country's authority thins, both in the letter of the law and in the physical control they could feasibly pull off.

The United Nations Convention on the Law of the Sea hoped to establish a "legal order for the seas and oceans." Over 160 countries have signed UNCLOS. Reading the two-hundred-page document, it seems as though the ocean is under control. There are rules and regulations against polluting, overfishing, and human trafficking. We support monitoring the safety of ships and protecting the marine environment. But on the water, it's a different story. Up to 26 billion metric tons of fish is landed illegally each year, and over 30 percent of fish stocks are past their breaking point. Shipping companies are notorious for off-loading oily waste at sea; cruise ships have the same reputation with sewage. Nations neglect to crack down on destructive fishing practices or they overlook barely seaworthy vessels that belch smoke, leak sewage, or employ slave labor. Some nations are building islands in the middle of the sea in order to extend their jurisdictions and extract even more resources from the ocean.

Each nation controls their sea as they see fit without acknowledging a rather obvious characteristic about water: it moves—constantly. "One country could have the best conservation initiatives and the country bordering it could have no best practice in place," Chloé Dubois, the leader of the plastic-collecting expedition, told me in MuQwin Peninsula. "And then the whole thing would be kiboshed." Species that move with the water tend to suffer the most, like Atlantic bluefin tuna, the great travellers of the ocean. These beautiful, athletic fish cross the ocean and swim in and out of jurisdictions, some of which have fishery management practices and others which don't. The result is a fish that has fallen off a steep cliff in terms of population numbers—a catastrophic decline.

Whether it's plastic or bluefin tuna, what's off your shores today could be off someone else's tomorrow. We cannot own one part of the sea and take no responsibility for the rest, because the rest of the ocean is not some small amount. An estimated 64 percent of the ocean falls within what's called the high seas: half of the planet's surface that is ruled by no country and governed by international laws that are so flagrantly flouted you can watch them being broken on reality TV. Anti-whaling laws have been in effect since 1986, but on seven seasons of *Whale Wars*, the hard-line animal-rights group Sea Shepherd filmed Japanese ships harvesting whales for research purposes, or so the Japanese claimed. In 2019, Japan started to hunt whales openly again after withdrawing from the International Whaling Commission.

Why does the freedom of the sea persist when it actually looks more like chaos? Perhaps it's because freedom seems to fit with our long-running historical and cultural concepts of the ocean. Perhaps it's because we never had the ability to control the ocean, so we call this lack of control "freedom" instead. The problem with freedom is that everyone defines it differently. I saw this borne out in the ocean stories of sailors, migrants, and workers. For offshore sailors, freedom at sea appeared in its purest, most romantic form. The average person could escape the constricting rules of land and sail across the largest ocean on Earth. For cruise ship passengers, freedom looked like a cheap vacation, an escape from the stress and routine of land. For cruise ships, freedom meant avoiding the rules and regulations of the lands where they are headquartered. Occasionally, I even met a disadvantaged person who turned this concept of freedom to their advantage. But that was rare, and in those cases, freedom at sea looked so dangerous it seemed more like desperation. Too often the concept of freedom was corrupted into a Wild West brand of law breaking.

Safeguarding so vast and unpredictable a territory, as well as the people and animals who cross it, is beyond the scope and resources of any one nation. It requires a collective effort unlike anything the world has ever mustered. When no one takes responsibility, when we "still respond automatically to the shibboleth of the 'freedom of the seas,'" as ecologist Garrett Hardin wrote in his essay "The Tragedy of the Commons," the ocean is destined for decline.

The ocean as we know it today has perhaps a decade left before catastrophic changes take hold, such as a massive decline in wild fisheries and the spread of oxygen-depleted dead zones at sea. As the ocean absorbs more carbon dioxide and acidifies, we change its chemistry. As phytoplankton bloom in the poles and bust in the tropics, we change its color. Some species are doomed to disappear; others to flourish. Many scientists believe that our future is already sealed, but what's apparent is that we are heading into a new and more turbulent relationship with the ocean.

We learn in school that there are seven oceans, as if we could move on to another when one is all used up, but there is only one great interconnected ocean, constantly circulating in and out of seven major basins, and in and out of billions of lives. The water that is in your life today will circle around to someone new tomorrow, working through currents, across the planet, through generations, time, and space. If we can use it, heal it, and pass it along in a healthier state to someone new, the ocean still has a chance for renewal and redemption.

As I spoke to people about why they are on the water, I witnessed our understanding of the ocean widen as the years passed. During a time of rising sea levels and catastrophic floods, of more frequent and intense hurricanes and declining fish stocks, extinctions, and pollution, the ocean is no longer only an "environmental" problem. We're seeing how the ocean connects to food, transportation, infrastructure, human rights, mental health, technology, and economics. The list goes on. The ocean shows us how deeply we are a part of our world.

By the sea, we live in the moment. We feel our fear and happiness for the future even more keenly. We cluster into groups and defend our turf. Across a world of differences, the ocean makes us more alike. We might never feel that conscious connection with the sea, but it's there. For our survival, as well as the ocean's, it's time to look past the surface and discover the connections down below.

THE IMPERILED OCEAN

ONE

CAPTURING THE WATER WORLD

"CUT!"

The action on the set of the top-secret Kanye West music video ground to a halt once again. As the sunlight faded from the sky the shoot in southern L.A. geared up for a long night ahead. Producers and their assistants, grips and gaffers, lighting techs, teamsters and talent all milled around a water tank, seventy-five people in total waiting for a beautiful young model to ready herself for another dip. She was having trouble holding her breath long enough to nail the shot. In the distance, the lights of Long Beach Airport spread through the darkening sky. The woman breathed in and sank beneath the surface one more time.

"Roll camera!" the director called. The 4,000-watt lights blazed, swirling like quicksilver over the surface of the pool. Floating underwater, cinematographer Pete Romano steadied the camera on the woman. His frame was sheathed neck to toe in a wetsuit, his bald head wrapped in goggles, his camera inside a watertight housing. Before he could hit Record, the woman surfaced. Five, maybe ten seconds had passed. She couldn't stay underwater for even a standard thirty seconds, and Pete couldn't get the length of shot he needed. Something else was going on with her, Pete realized. He had seen this panicked reaction on water sets before.

When a film moves below water, a change in command usually happens behind the lens, and someone like Pete, who is specially trained to film in water, takes control. As one of Hollywood's top underwater cinematographers, he's filmed hundreds, if not thousands, of people in water. Over the decades, he's become responsible for how millions of people experience the ocean on-screen. Pete is a sort of translator. He shows us what a vast body of water looks and feels and moves like. His fingerprints are all over the last thirty years of underwater filmmaking: from James Cameron's *The Abyss* to *Free Willy* to *The Life Aquatic with Steve Zissou*. What his credits have in common are big-budget productions that gross millions of dollars at the box office. And water.

On the job, Pete is thinking about the trim of his camera and frame composition, but he's also thinking about his buoyancy and keeping his upper body steady as he fins through the water. His is a tricky, obscure job that calls for the body of an athlete and the eye of an artist. Around his waist, he straps a twelve-pound weight belt and a knife; in his hands he guides a perfectly weighted camera inside its underwater housing. All this equipment weighs around fifty pounds and takes hours to assemble and prepare. His days are long, filled with shoots like this one, and his packed schedule is a testament to the ocean's enduring storytelling appeal. From *The Odyssey* to Shakespeare's *The Tempest* right up to Herman Melville's *Moby-Dick* and James Cameron's film *Titanic,* the ocean is where we struggle, die, fall in love, or transform into someone new. On the big screen, water hits key emotional chords. The ocean is cast as heaven or hell, redemption or punishment, escape or trap, inspiration or nightmare.

Rarely does it play anything in between. Water takes a polarizing turn on-screen, and it divides actors offscreen as well. They tend to either excel or flounder, and if the actors can't perform, it's up to Pete to figure out some sleight-of-hand solution on the spot.

Underwater, command over something as familiar and intimate as one's own body becomes challenging. We can't speak. We can't hear very well or see very far. Before shoots, Pete often encourages performers to practice in water, but that advice is not always heeded. People tend to dismiss water because it's so familiar: in a bathtub, a glass, a day at the beach. But when we're surrounded by the stuff, when we're in over our heads and feel that swirling loss of control, it stirs up a primal reaction, both on-screen and off.

Considering our evolutionary background, humans should feel comfortable in the ocean. After all, it's where we came from. All life on Earth owes a deep, incontrovertible debt to water. Aquatic animals, like sponges, are millions of years older than terrestrial ones, like apes. We still carry this background in our biology. In the womb, we develop aquatic features before terrestrial ones, like the two-chambered heart, fin-like limbs, and slits that resemble fish gills. As infants, we reflexively know how to breaststroke before we can walk. We are all endowed with the mammalian diving reflex: an evolutionary relic triggered by water that allows us to survive longer without air.

The diving reflex is particularly strong in infants. With nothing more than a puff of air on their face, newborns involuntarily hold their breath and brace for water. They can last forty seconds longer than an average adult. When we learn to walk, we lose our longer breath-holds. On land, typical adults can hold their breath for thirty seconds. As the seconds tick past, acidifying carbon dioxide accumulates in the blood, and the feeling of pressure in their lungs starts to build, as does a sense of panic. They eventually cave and take a gasping breath of air. In water, the same series of events unfolds, but slower. When cold water triggers the diving reflex, we involuntarily hold our breath; the heart rate slows and the body goes into triage mode, redirecting blood from the fingertips and toes to the heart, brain, and vital organs at the core. Our need to breathe slows and our body knows to pace itself for that next breath. Without training, a typical adult can hold her breath for up to a minute in water—double the rate she can on land.

Despite our biological beginnings in water, humans evolved on land, and our features are shaped accordingly. Water dampens our senses. Our lungs seem to tighten after only a minute without air. Perhaps the ocean's inaccessibility to us is what makes the depths all the more alluring. We want to see the underwater world that should be off-limits to us. Some of us will invest in expensive gear and training to experience it personally. Thanks to people like Pete, the rest of us can see what's down there from the comfort of our living rooms.

~

In the late 1800s, the French diver-biologist-photographer Louis Boutan took the first underwater photograph meant purely for pleasure. Boutan's image is a haunting black-and-white portrait of a diver, sheathed in brass, surrounded by a dark, impenetrable ocean. By the diver's feet is waving sea grass; bubbles stream up and away from his helmet to the dimly lit surface above. The gritty photograph is far removed from the high-definition footage Pete Romano shoots today, but it captures something elemental about the ocean. This deep, dark place was not meant for humans.

The ocean has a long history of rugged manly pursuits: sailors, merchants, whalers, pirates. For centuries, the ocean came with a warning, "Here be dragons," scrawled ominously across early maps. The Roman writer Pliny the Elder speculated that all land animals had an ocean equivalent, so early imaginings of marine creatures were often terrestrial animals crossed with tails or fins: lions with fish tails, sea monsters with rhino horns, rams and horses that swam. Sailors brought back stories of sea serpents and mermaids glimpsed in the waves. Some of these stories might even have been the early warning signs of marine debris, in the form of discarded fishing nets and lines dragging down whales and sea turtles. Many encounters with the ocean's unknown often veered off into the darker regions of the imagination. In Greek mythology, beautiful sirens lured seamen to watery deaths; in Norse legends, the gigantic kraken swamped ships; and on early maps sea serpents are shown feasting on ships and their crew. These fabulous tales are not so different from the stories told on the silver screen. Drawing on humanity's deep-rooted fear and fascination with the ocean has always made for great movie fodder.

Although Pete calls the water his office, his base is a twenty-six-thousand-square-foot warehouse in an industrial area near the Los Angeles International Airport, a one-stop shop for filming the underwater world with the most advanced gear. It's also a mini-museum of our long-running fascination with diving the deep and recording what we encounter. Near his desk sits an antique diving helmet similar to the one worn by the diver in Boutan's photograph. In another room of the warehouse stands a heavy brass diving suit from the Second World War that weighs about two hundred pounds fully rigged. A knife is sheathed at the waist, and it bears little resemblance to contemporary scuba suits. This underwater armor was the first inspiration for early space suits. The similitude between the ocean and outer space is not such a stretch. Astronauts train in a vast swimming pool that NASA dubs the Neutral Buoyancy Laboratory. "It reminds me of a crusader," Pete said as he touched the metal and chuckled. "When men were men," he added sardonically. Pete's voice was hoarse, his bones creaky, from wrapping the late-night Kanye West shoot two nights earlier. With his bald head and gravelly voice, he has the look and intensity of the character actor J. K. Simmons, but with a languorous Boston accent.

The hallways of Pete's office are lined with film posters commemorating underwater pioneers like Jacques Cousteau and Hans and Lotte Hass, an Austrian director-couple who filmed *Adventures in the Red Sea* in 1951. Lotte Hass nearly drowned on set. "You cannot ignore the people who did it before us," Pete said. "The people who had no monitor or no viewfinder, they just got down there and duked it out."

Then came the posters for the more outrageous underwater films, most of them only dimly remembered today. The Tarzan film *Dive of Death*, *Hell Raiders of the Deep*, *Jaws of Death*, and *Manfish* are just a few of the titles. Each poster looks like a cover for a pulp paperback with some combination of brawny men, scantily clad women, and fearsome ocean creatures. In the early 20th century, the major ocean institutions like Woods Hole in Massachusetts and Scripps in San Diego were still in their fledgling years. For the masses, the ocean was a blank slate, giving directors free rein to imagine the most magical, dangerous, and ludicrous scenarios. One poster for the film *Zombies of Moratau*, circa 1954, added an implausible twist of zombies walking the seafloor. But half a century later, when we have far more scientific knowledge about what actually lives and breathes beneath the waves, we're still making films about seafloor-strolling zombies, like in *Pirates of the Caribbean*. The ocean is often cast as the innate instinctive id to land's critical and self-contained superego. On-screen, the water allows us creative license to imagine our most irrational fears about life in the watery regions of the nether world.

"It's just turned out to be some silly thing," Pete said nonchalantly as we strolled through his warehouse: shelves upon shelves of lights, radios, underwater speakers and tripods, video monitors, and remote-controlled camera housings. The field of aquatic film gear developed in tandem with scuba gear throughout the 20th century.

To physically go and explore the underwater world in comfort, humans need a lot of equipment. Just as aristocrats who had time and money on their hands pioneered the sport of offshore sailing, those who could pay for the cutting-edge equipment that would push them beyond human limits were the first to explore the depths. Working-class men wouldn't have had the resources to explore the ocean for exploration's sake.

Divers once walked the bottom of the seafloor with weights on their feet, dragging a hose that pumped air into helmets. It was a gentlemanly way to explore the ocean, walking upright rather than swimming like a fish, forcing the aquatic to conform to the terrestrial. But just as Icarus built wings to fly like a bird, we stopped walking the ocean floor and instead developed goggles and fins that more closely approximate a marine animal's experience. For about as long as we've had the gear to stay underwater longer than a few minutes, we've taken cameras down to record what we see.

Pete's life in the niche world of underwater film began by chance when he was in his early twenties. He went for drinks with a friend who had just been rejected from the Navy's combat camera unit for his poor eyesight. Pete had decent eyesight and a boring job he wanted to escape. In one of those impulse decisions that end up changing a life forever, he decided to apply.

"I really fell in love with the frame. I really fell in love with water," he said about his time with the Navy, travelling to the Philippines, Taiwan and Hawaii to shoot underwater units in action. After finishing up with the Navy, Pete tried to transition back to civilian life by looking for a job in underwater film. He called up a cameraman he knew who was looking to become the American Jacques Cousteau. The man didn't hire him, but he told Pete that a machining background would make him an ideal hire. This was because the underwater gear for rent at the time was appallingly inadequate, according to Pete. The cases (also known as underwater housings) had springs inside that had to be hand-wound; the camera focus was mostly guesswork. One popular camera came with a slot in the back to stuff with half a roll of paper towel to absorb moisture and leaks.

The ability to repair a housing or even make a better one would be a surefire way to advance in the field.

When Louis Boutan captured that early underwater photo, he inserted a flat piece of glass on the front of a waterproof case. The problem with this jerry-rigged invention is that flat glass, also called a flat port, records the ocean through two elements: the air on the inside of the camera's housing and the water on the outside. Water is eight hundred times denser than air, and light radically slows down underwater. The flat port produces a distorted and magnified image. This looks closer to the human experience of seeing underwater through a diving mask, but we want to suspend our disbelief, to see the ocean world in sharp detail and vibrant colors, the way we do above water. In the 1930s, a hemispherical dome port corrected for the refractive distortion of filming underwater. The curved lens allows light to pass through without refraction and creates what's called a "virtual image." This optical illusion shifts the underwater world closer to the lens the same way a rear-view mirror makes objects appear closer than they truly are. When Pete films with a standard dome port, he's not actually filming the flesh-and-blood person but the virtual image that seems to appear about sixteen inches in front of his lens. It's a bit of magic, learning to film in water, an element that seems simultaneously familiar and strange.

On an underwater shoot, Pete places more and better lighting above the tank to correct for what's naturally a dark place. The deeper down the camera goes, the more colors leak away. First the long-waved reds disappear, then the oranges, yellows, greens, until the deepening and darkening shades of blue, violet, and finally black take over. (This rule applies to the colors of fish, too. There are few blue fish in the midwaters of the sea because that shade persists longest there and predators would easily spot them. Red, orange, and yellow look like attention-grabbing traffic signals above water, but below, brightly colored fish fade to black the quickest.)

In the early days of film, all these filming challenges made the ocean look like a murky place, drained of light and color. *Creature from the Black Lagoon*, filmed in 1954 and possibly the only early underwater film that people might still remember today, is all gray-scale water with the occasional silver flicker of a bubble. Two years later, French filmmaker Jacques Cousteau awakened the world to the vibrant color of the marine world in his landmark film *The Silent World*. Swarms of electric yellow tangs flickered across the screen. Red groupers peered out grumpily from between coral. Luminous pink anemone tentacles waved, and a giant green turtle finned past. Before *The Silent World*, few people knew what life and color existed down below. The impact was instantaneous. Now nearly every beach resort sells snorkelling and diving packages to fit a vacationer's schedule.

After completing a two-year machining course at San Diego Community College, Pete earned extra cash diving with a boat called *The Bottom Scratchers*, named after a legendary group of San Diego skin divers. "I dove for food, because I couldn't afford to buy food, or I dove to take pictures," he remembered. On *The Bottom Scratchers*, Pete met Lamar Boren, a titan of underwater cinematography, who shot *Sea Hunt*, a popular black-and-white TV show during the 1950s. *Sea Hunt* followed the exploits of a former Navy frogman turned freelance scuba diver, played by Lloyd Bridges. The gravelly voiced Bridges was like an underwater detective, always on the case of buried treasure or rescuing a damsel in distress. Boren became an inspiration and hero for Pete, who quickly realized that he didn't have the patience for underwater documentaries and set his sights on Hollywood instead. Pete still reveres the work of Lamar Boren today, particularly a five-minute underwater fight scene Boren shot in the 1965 Bond movie *Thunderball* that he calls a landmark for the field.

Pete's first credit as a camera operator was on *Jaws: The Revenge* in 1987. His latest credit won't appear until a year from now, when one of the half dozen movies he shoots or consults on every year is released. He's become the go-to specialty guy flown in to nail a shot of Sandra Bullock struggling to remove her wrist ties as she sinks through water in *Speed 2: Cruise Control* or Nicolas Cage drowning his daughter's kidnapper in *Stolen*, a thriller in the vein of Liam Neeson's *Taken*. Typically, Pete spends a day or two on a set before moving on to the next gig. His contribution to the over 150 films on his CV can pass by in seconds, but they form the crucial backbone of films like *Flipper*, *Free Willy*, various installments of *Mission Impossible*, *Inception*, and *The Life Aquatic with Steve Zissou*.

Forty-five years after that impulse decision to join the Navy, his company, Hydroflex, employs around ten people. Young assistants in adventure clothing wheel pushcarts up and down the aisles of his warehouse, cleaning out splash bags and readying new ones for rentals or packing cameras into hard-plastic cases to ship to sets around the world. A machinist works the lathe, repairing one of the Hydroflex housings that took underwater cinematography to a new level. This is the work that Pete used to do himself. "When I'm on set and I hear Steven Spielberg go, 'Okay, let's bring the Hydroflex over here,' I like that."

All this goes to say that Pete is a big fish in Hollywood's underwater film world. Katie Rowe, a water-stunt coordinator from Long Beach, estimates that the field has around a hundred regulars in the United States. That includes all departments, from cinematographers to marine grips, lighting, stunts, set decorators, boat operators, medics, and mermaids. "We all work together over and over. We cover for each other and work for each other and fill in for each other all the time," said Katie. About half her work involves water-based stunts, either training performers or performing them herself. "When I normally get a call from production that they want to do some water thing, some of my first questions are who's your camera operator? Who's your underwater grip crew?" She listed off a few of the names of underwater cinematographers she likes to hear, Pete Romano's among them. Her worst days on set are when the production hires a friend who scuba dives but has never shot underwater sequences before. Everybody's got to learn, she allowed, but those sets often turn into disasters. Waiting in a cold tank while someone figures out how to wrangle lights, set up equipment, and handle a camera underwater is not fun or safe.

Pete almost never shoots in the ocean anymore. "Tank," he answered immediately when asked his preferred filming location. "Otherwise you're at the mercy of the water. The sea conditions, the sky conditions, water conditions—all of that. You're in an organic moment." Many of the box-office-busting underwater moments you know and love were likely shot in a tank. For *Life Aquatic*, Pete shot a Cousteau-inspired scene of Steve Zissou and his team diving into the ocean depths and holding sticks of dynamite to light the way. That was actually shot in a eight-foot pool, and Pete tilted the camera to hide the walls.

A tank allows Pete to rule out the wild elements of the ocean, to control the water inside the frame, where even the smallest smidge can ruin a shot. Crystal-clear water is a must, for instance, in up-close shots or scenes with swappable blue-screen backgrounds. Big-budget Hollywood productions aren't the only ones that opt for controlled conditions where they can adjust the lighting or cheat the angle. BBC's *Blue Planet*, that bastion of in situ nature documentaries, has come under fire for representing aquarium and zoo animals as "wild." *Blue Planet* defends the practice; some animals are so microscopic or live in such deep and dark marine settings that they would be impossible to film otherwise.

It could also be that fewer pristine spaces with abundant animal life and clear shooting conditions are available today. Perhaps drifting specks of algae or microscopic plastic would interrupt the narrative with a little too much reality, anyway. When it comes to capturing nature, our beauty standards are high and often divorced from the real world. In one study from the University of Queensland, citizen participants equated fluorescent coral reefs with health and abundance, but the bright color was actually an early sign of stress and bleaching. We're learning about the ocean through a screen, which has led to an ever more idealized and sensationalized understanding of the sea. Pete's work still has a foothold in reality because he films live-action people in real water. That collision between humans and the elements will always yield new creative surprises.

Even in a tank, Pete can't rule out the unpredictable side of water. There's always some variability that can't quite be contained, like actors and models who find performing in water overwhelming. Scarlett Johansson is one of the A-listers who is good in water (she's a real pro, Pete says). Never mind who's bad—Pete didn't survive this long in Hollywood by trash-talking. Some sets pay for a water-stunt coordinator, like Katie Rowe, a former competitive swimmer who has spent her whole life working in water. With particularly tricky or technical scenes, she has asked for up to two months to train the talent before the shoot. "It's not generally just swimming," she explains. "It's situations that are your worst nightmare, like a plane crashes into the water, and you're trapped in it. Or you're being attacked by a shark, or you're in a boat that's flipped upside down. Or you're in a room that's sinking. They love that one."

She finds that children often adapt the quickest to working in water. "To them, it's just another day playing at the pool." But too often, experienced performers assume that if they like the beach or swim laps at the pool, they'll perform well underwater. "People always say 'oh, I can swim,' and I say, 'yeah, but can you drown?" Performers don't necessarily factor in the nightmarish scenarios they might be placed in, or doing repeat dives, running on little oxygen and sometimes even staying underwater for long periods of time while Katie passes a regulator between takes. The walls of the tank are often painted black, so there's no frame of reference. The shoot can drag on for hours.

"A lot of people are claustrophobic," said Pete, before adding, generously, that filming in a tank is also a naturally claustrophobic moment. Behind that feeling of claustrophobia can lie a fear of drowning—the ultimate loss of control in the water. Drowning is often portrayed as a peaceful way to die; perhaps we see symmetry dying in the same liquid that gave birth to all life on Earth. But the feeling is not a pleasant one. Suffocation stirs up an ugly, grasping, elemental panic, whether we're trapped underwater or simply choking on a piece of bread. People who survive a near drowning say they see darkness close in on all sides, like a "camera aperture stopping down," as journalist Sebastian Junger put it. In his surf memoir *Barbarian Days*, William Finnegan wrote about the worst waves, the ones that held him down longer than he expected, and how "that last extra kick, or two, or three, still without reaching the surface, made the desperation for air, the spasm in the throat, feel suddenly like a sob or a stifled scream. Fighting the reflex that wanted to suck water into the lungs was nasty, frantic."

The sport of free diving prepares the body and mind for voluntary drowning by learning to control what is normally a terrifying experience. Katie Rowe, for instance, can hold her breath an astounding four minutes and thirty-eight seconds underwater. Competitive free divers reach mind-blowing depths on a single breath of air and often say they feel a sort of transcendence when they touch the limit of what their bodies can do. Some believe that our bodies are capable of much more if only we learn to tap into the diving reflex, that safety feature that every mammal is born with.

On the Kanye West shoot, after the model tried again and failed to hold her breath, Pete removed his breathing regulator and swam to the edge of the tank to parley with the music video's director, Hype Williams, king of '90s rap, hip-hop, and R&B music videos. Pete tried to convince Hype that the shoot just wasn't going to work with this model. But Hype was adamant: the girl stayed in the picture.

After seeing performers panic in water dozens of times over decades of shoots, Pete has developed a pet theory for why it happens. Water can make us feel out of control, he explained. This feeling must be especially difficult for performers of all stripes who above water wield their body with precision. In situations like these, Pete tries to ground performers by giving them something to hold onto in a new and unfamiliar setting. It can be as simple as a twelve-step ladder. Pete had been using one to hold himself in place by jamming his flippers into the steps. Now it seemed like the model needed the stability more than he did.

"I want you to stand on this," he said, placing the ladder in front of the model. "When it's time to go down, I want you to walk down the ladder." He demonstrated how to hold the top rung and swim forward toward his lens. This way, he could fake the image of her drifting at sea without her ever having to take her hand off the rung. He sank into the pool and waited.

Over the underwater speakers, the director's voice floated to his ears: "Roll camera!" The model walked down the ladder, gripped the top stair, and floated forward. She held her breath; Pete moved in for the shot. Five seconds passed, ten, twelve. The woman twisted and turned for the camera, hair and clothing swirling, fifteen seconds, twenty seconds. She was out of the water and back up for air. Pete got the shot.

"Okay, great!" Hype called out. Now on to another shot, another angle, something new. That night, the shoot lasted another seven hours. At 2:00 A.M., Pete climbed out of the tank, his eyelids droopy, and packed up his dive gear.

In 2001, Pete and Hype had shot another beautiful woman swimming through a dreamy waterscape, on the set of the video for Aaliyah's "Rock the Boat." The song is about making love, using thinly veiled nautical and watery references. Aaliyah tells her lover to coast, to go fast, to change directions, and, obviously, to rock the boat, all while dancing on a catamaran and rolling around on a beach. The final images of the video show Aaliyah immersed in the sea. This is where Pete came in. In a pool at Florida State University, Pete swam beneath Aaliyah, shooting up toward the surface, while she twisted and turned through the frame, looking down from above. Her flowing black clothing and black hair look stunning against a cloudy backdrop, created by a billowing smoke machine on the pool deck.

These shoots are on the opposite end of the spectrum from the nightmarish escape scenes that Katie Rowe trains talent for. They capture the beautiful, inspirational, sexy side of water. The image of a beautiful woman underwater, naked or draped in flowing clothing, is a classic, stretching back to the Greek and Roman gods of love, Aphrodite and Venus, born from sea foam, right through to Arthurian poetry, Shakespeare, Victorian paintings, and modern-day photography and music videos. The image also has a darker underside. Water transforms women into goddesses, gravity-defying nymphs who float free from earthly restraints, but it can quickly turn on its head, showing them as dead, drowned, trapped or somehow tainted by the sea.

Throughout literature and art, a drowned woman is often shorthand for a heartbroken or fallen woman. In *Hamlet*, the rejected and miserable Ophelia drowns herself in a river. In Tennyson's "The Lady of Shalott," the pure but isolated young lady descends from her tower to taste the real world and ends up drifting downstream in a doomed pursuit of earthly pleasure. In George Eliot's *The Mill on the Floss*, the main female character, suspected of having lost her virtue, dies in a flood.

In the music videos of contemporary pop songs, the image recurs again and again, from Selena Gomez's "Come and Get It" to Rihanna's "Roulette." Beyoncé's music videos in particular rely on water to communicate the singer's inner struggle or joy, culminating in the coup de grâce of "Hold Up." For a minute and a half (an impressive amount of time for an underwater sequence in a music video), Beyoncé writhes in a lavish and flooded bedroom, tormented by her husband's infidelity.

All this imagery suggests an earlier time when women and water were not meant to mix, when even bringing a woman on board a ship was considered bad luck that doomed the crew to an ill-fated journey. Throughout history, men dominated the sea, crossing it, exploring it, battling it. On-screen, men on boats typically fight the elements in films like *All Is Lost* or *Master and Commander.* Women are more likely to have fallen overboard and need rescue, or else they're already lost causes.

According to Katie Rowe, abstract dreamy sequences can be easier to shoot than plane crashes or sinking rooms, particularly when the production hires synchronized swimmers or stunt doubles. (The ease of an underwater shoot often comes down to money invested.) But still, they come with their own challenges. "Production tends to love things that look beautiful in the water," she says, "like long, flowing skirts, things that kind of swirl. Ribbons, when you're thrashing around, tend to wrap around you like a snake and try to sink you."

On the "Rock the Boat" set, Aaliyah swam through a tropical paradise that seemed to hold no macabre overtones. The ocean and the water were all about sexual pleasure and exploration. After Pete shot the final frames of the star swimming in a pool, the shoot moved to the Bahamas where they would shoot the rest of the music video. Aaliyah's plane later crashed on her return to the United States, killing everyone on board.

After Aaliyah's death, the meaning of the footage Pete shot, and indeed the whole music video, shifted. At first, this beautiful woman swimming in the pool was playful and sexy and beautiful, and ended with her swimming off into a happy ending. After her death, the music video seemed to signal a more permanent farewell. In the final frames, it's hard not to see Aaliyah looking down from the heavenly beyond, the backdrop infused with clouds and rays of light cutting in from above. She reaches out to the lens, her hair billowing and swaying like an angel, and then she swims to the surface. That footage later came to haunt Pete, as though he had seen her death foreshadowed in the frame. Or maybe it's that the image of a woman drifting in water is always precariously balanced between sex and death, two forces eternally linked in women's lives.

Water is the world's universal solvent, with a nimble quality unlike any other liquid on Earth. It transforms quickly among ice, fog, and liquid. It absorbs the light. It reflects the world around it, creating a pleasing symmetry. It slows down movement, elongates a graceful stroke. There's something deeply hypnotic about the way water moves. It feels inexplicably natural and pure, evoking the flicker of a flame or the wavering billow of smoke. Many of the touchstone sequences that Pete films play with water's natural shape-shifting qualities.

In a tank, Pete can tease out what he wants in a controlled setting. The light, the angle, the bubbles, the spray all become the perfect foil for a character's quest. In *Edge of Tomorrow*, when Tom Cruise nearly drowns in a classic all-is-lost moment, Pete aims the camera down into the depths of the tank where the color leaches into black. In a hopeful redemption scene later on, the hero swims back to life, and Pete shoots toward the surface, creating a background intercut with rays of sunlight. In *Across the Universe*, two lovers embrace in a swirling stream of passion, bubbles escaping from between pressed-together lips.

Although Pete's work demands controlling all the variables in and around the frame, making sure there's enough light, the right exposure, and adjusting constantly to capture the most evocative angle, a moment sometimes comes when he can only stand back and record water doing what it does best. Some of Pete's best images place water at center stage, letting it play the unpredictable creative force. In the 2000 film *Pearl Harbor*, Pete filmed underwater scenes of the SS *Arizona* sinking. Right in the middle of a montage of death and destruction, as waves of Japanese kamikaze fighters rained down hell-fire on the Hawaii army base, a stunning underwater image flashed past. Amid the legs of drowning sailors, a bloodied American flag unfurled in the water, backlit by a soft, rippling light.

"The shoot was planned," Pete remembered. "We had the lights set up. We knew what we wanted to get, but the way the flag unfurled . . . It was just magic." He still got chills thinking about that image, about something in the way the light hit the water and lit up the flag. That half-second of screen time managed to capture the whole thrust of a patriotic film: America, cut down by destruction, would surface again. It isn't an accident that these underwater moments are crucial plot points or thematic undergirding for an entire film.

Capturing these evocative moments in water is not cheap. To make the production's budget, an underwater scene has to be absolutely essential to the story line or it's cut from the script. "Water work tends to be slow, expensive, and there's a lot more safety required, a lot more specialized moving parts," said Katie Rowe. "A lot of times when writers write in this fantastic water scene or just a little water scene, it will get eliminated because it tends to be a small part of the story."

The most massively over-budget films in movie history all took place on the ocean or involved a lot of water: *Water World*, *Titanic*, *Jaws*, and both versions of *Twenty Thousand Leagues Under the Sea* broke records in cost overruns. After Steven Spielberg finished *Jaws*, he came up with the industry rule that water triples everything on set. What takes one hour, one hundred dollars, and one person on land, takes three hours, three hundred dollars, and three people on water. (Pete Romano says this is exaggerated. According to him, overruns on *Jaws* were more likely due to poor underwater equipment at the time and preparation.)

Tank One Studios, where Pete filmed scenes for the Kanye West video, is one of the few water tanks of its size in Hollywood, and it has a booking fee that would cut out smaller student productions straightaway. Before setting foot on the Tank One Studios lot, a production has to purchase $5 million in liability insurance. The tank is situated on a quiet industrial road lined with a machining shop and tree-mulching yards next door, a strip club across the street. No signs outside read TANK ONE STUDIOS, nothing at all indicates that multimillion-dollar shoots go down behind the chain-link fence, only a huge American flag flapping in the wind. The location is low-key for a reason. When word got out that Tom Cruise was filming *Edge of Tomorrow* at Tank One, the strippers across the street climbed on the chainlink fence, angling for an intro or at least a glimpse of Tom.

Filming in water involves many other costs. Heating Tank One's 233,000-gallon tank to a bathwater ninety degrees costs $3,750 in summer. A one- to two-day shoot typically spends $3,000 on propane alone. According to Katie Rowe, some producers try to cut costs by skimping on heating, and that can make for a long, cold day for the talent. Water shoots also require specialty equipment, a water-safety team, and water-quality checks that need to be completed every few hours. The tank needs to be vacuumed before each shoot and the water is prepped, too. A production might want crystal-clear water or a more mysterious cloudy look, in which case they pour milk into the filtration system. Before each shoot, the Tank One crew informs the production that they can test the clarity by flipping a quarter into the pool. Sometimes the production takes them up on the challenge and, if they can see George Washington's head through sixteen feet of water, the tank is good to go. Music videos have gummed up Tank One's filtration system with glitter; action movies have dunked cars that leaked gasoline and grime into the pool. A full drain and refill is an additional charge. But once the cameras roll, this water tank transforms into the Everglades, or a mermaid's lagoon, or the bottom of the ocean. Outside the frame is a dusty industrial lot scattered with propane tanks, oxygen cylinders, and shipping containers.

Along the inside rim of the tank are railings where scaffolding rises and green screens can be hung in place. In recent years, computer-generated imagery and live action often work in tandem. In the 2015 film *San Andreas*, the Hoover Dam crumbles after a massive earthquake and releases a watery deluge across California. Using ever-improving software, like Houdini, Maya, or the open-access tool Blender, animators set start and end points and obstacles for the deluge; tinker with the water's velocity, direction, viscosity, and flow; texture the foam and bubble, the waves and splashes, to make it look as realistic as possible. Then they bake all the elements together and let that digital creation loose in the modelled environment. These digitally rendered disasters look believable to the average viewer. When I watched *San Andreas*, I felt no uncanny-valley response to all the digital water cascading toward a seismologist played by Paul Giamatti. It looked real enough and had the desired effect of increasing my heart rate and compelling me to shout at Giamatti to get out of the way, to run faster already.

But, according to Pete, there's a good reason why CGI hasn't entirely replaced underwater film yet. Not long before the Kanye West shoot, Pete attended a screen test for an actor swimming in computer-generated water. "It was supposed to be a mermaid going through the water, a physical being swimming," he explained. "It was such a linear move that it just looked like a joke, like it was on a wire. It could be a parody, but that's about it." He left the screen test feeling reassured about his future in the business.

Visual effects animator Will Wallace, who contributed to *San Andreas*'s fluid dynamics, as water is known in the visual effects world, admitted that water is still the most complex element to portray digitally. Underwater is even trickier. "If you have a guy running through a pond, it's easy enough to get a group of particles to react like water would," he explains. But the water world is a total environment and demands another approach entirely. Instead of animating a watery object or group of particles, animators layer atmospheric elements, like bubbles, light, shadow, or wavering hair, until the desired look and feel is achieved. Nailing that believability is also trickier, perhaps, because everyone knows how water moves when you swim through it, how your body movements slow down or bubbles move across your face as you breathe. It's not easy to get a realistic look, said Wallace, but, technically, it's achievable. If someone doesn't believe the effect, it's probably due to limiting factors like time, money, or talent, but not technology. For Pete, who has spent years capturing water's immersive, 360-degree experience, his specially trained eye might never buy the digital simulation. After *San Andreas*'s release, seismologists and structural engineers spoke with news outlets about the film's scientific inaccuracies. "It looked as if there were a series of internal explosions within the dam, and that's not a phenomenon we would expect, or how we'd expect the damage to occur," a professor of earthquake engineering told the *Guardian*. The expert will always discern the stumbling points where fantasy is pushed too far. The rest of us are probably becoming or are already comfortable with the simulation.

More and more films are relying on CGI to do the heavy lifting when it comes to creating these immersive moments on-screen. Disney's live-action *Jungle Book* was shot entirely on green screens in a Los Angeles warehouse. And yet considering the costs, the specialized knowledge, and the potential safety hazards that working with real water requires, it seems inevitable that as animation technology improves and becomes less labor- and data-hungry, it will one day entirely replace the real thing. According to one animator, the post-production credits for big-budget action movies already run far longer than those for the physical crew on set. In the future, instead of Pete Romano and Katie Rowe struggling with regulators or lights or difficult talent, there will be animators locked in a battle with computer programs. In a modern world where we rely on computer screens to navigate, connect, and capture the real world, this shift feels entirely expected. Of course we'll watch a simulation of a simulation of a story about our forever changing relationship with the sea.

The whole point of film is to lose ourselves in a world we might never experience in real life. This is where underwater cinematography first brought us when Louis Boutan took that first blurry photograph back in 1893. The journey to the bottom of the sea has been a beautiful one as underwater cinematographers have figured out how to capture light and color, inspire marine biologists, and create a field of specialized knowledge. All the tricks of the trade are designed to approximate that human feeling underwater, to bring us close to the beauty and the fear of the ocean world without the danger.

Pete Romano and the world of underwater film invite us to dream about the mystery of the ocean. His work is the most common representation on-screen, with its big stories of heroism and sacrifice, beauty and inspiration. Behind those stories is an entire life dedicated to working in water, to improving the gear, the equipment, the training, so we can look at the underwater world we crave. There's some value in this real-world collision between humans and the water. It's where unpredictable transformation happens and where water challenges us to push ourselves. Spending so much time immersed in the underwater world that Pete created on-screen, I felt myself craving the real thing again. I took that as a sign that he had done his job well.

TWO

SAILING THE SOUTH PACIFIC

Hone and spread your spirit till you yourself are a sail, whetted, translucent, broadside to the merest puff.

—Annie Dillard

Fiona McGlynn gripped the helm of her thirty-five-foot sailboat and focused her honey-slow, sleep-deprived thoughts on the sound of a wave shattering behind her. Her boat had already been knocked down once that night. But if she could angle the boat just right and surf straight down the powerful, slow-moving waves, without allowing them to crash on her stern, everything would be all right.

Four days earlier, on a clear September day, she and her partner, Robin Urquhart, had set sail from the west coast of Canada for a two-year journey across the South Pacific. As they sailed south toward their first stopover in San Francisco, they listened to the weather forecasts and paid special attention to their weather router: an expensive service where an expert recommends the best weather window for a boat's passage. They had splurged on the router specifically to ease their anxiety about unexpected conditions at sea. According to the router, the passage south should have been uneventful.

Between them, Fiona and Robin shared a few hundred nautical miles of sailing experience. This was their first long passage at sea on their 1979 Dufour sailboat. They'd spent the summer soaked in sweat and covered in itchy fiberglass, repairing *MonArk* in a shipyard: ripping up the deck, replacing the rigging, and installing new electronics.

The hard work was fun when it was fused with excitement for the ocean crossing ahead. That winter they would sail around Mexico's Sea of Cortez. In the spring—the ideal time to cross the Pacific—they would set off to the Marquesas. From there, they would island hop from Fiji to Samoa and to all the strings of scattered atolls and watery outposts across the South Pacific before ending up in Australia. If they were enjoying themselves, they might just keep going forever: happily ever after, sailing off into the horizon—if they made it that far. One of their biggest fears was that all their preparations would falter before they even made it to Mexico. Now, just a few days past the forty-ninth parallel, they were struggling to hold *MonArk* steady on a terrifying sea.

Off the coast of Coos Bay, Oregon, the fair weather started to shift. An ominous weather report from the National Oceanic and Atmospheric Association came over the satellite phone, warning local boaters to stay in port or take shelter immediately. For Fiona and Robin, neither were an option: they were sixty miles and at least a day's sail from land. "As the light fell, the wind and seas were building," Fiona remembered later. "Soon it was pitch black and the roar of white water filled our ears." An hour after the weather warning, the ocean quickly devolved into a chaotic cross-sea of high-powered waves and gale-force winds.

On land, a gale sways a whole tree, twigs break off and fly away. At sea, gale-force winds (about seventy kilometers an hour) can kick up twenty-foot waves. And it was the waves, not the wind, that was a problem that night: huge waves that rose so high they broke over themselves; messy waves that slammed into *MonArk* and sent her spinning across the surface. We often think of water as a forgiving liquid because we usually deal with it in such small quantities. But as anyone who has hauled the stuff over land or belly flopped into a pool will tell you, water is hard and heavy, and it hurts. Every few minutes thousands upon thousands of liters crashed into *MonArk's* port side as she heeled over at an eighty-degree angle. Today thirty-five feet is on the small side for an oceangoing boat. Fiona and Robin felt every crest, every drop, every slam, every valley, and more—like they were mountain biking without shocks. The waves were so huge they rendered the boat's self-steering wind vane useless. For the next twelve hours, Fiona and Robin hand-steered *MonArk* through the night, twisting the helm, trying to ride a straight shot down the shattering waves.

They had done everything in their power to avoid this spate of bad weather. What their weather router couldn't predict was the influence of land on local conditions, like the winds that accelerated up and over nearby Cape Blanco and crashed into their boat on the other side. It was safer not to attempt to sail to land in a hurry. In a storm, coast is more perilous for a boat than open sea. Land poses a risk you'll run aground and wreck the boat onshore; open ocean gives more room to make mistakes. The whole point of offshore sailing is leaving a safe harbor behind and facing down whatever the ocean sends your way. The average person, myself included, would consider this madness: sailing out to sea with full knowledge that bad weather might hit your boat. But the thousands of people like Fiona and Robin who do it every year, cruising from continent to continent without incident, no airports, planes, or reservations required, intrigue me.

"The sea, among its many qualities, is an unerring discoverer of weakness," notes Derek Lundy in *Godforsaken Sea*. Lundy means both the boat's weaknesses and the sailor's. After the summer rebuilding *MonArk* in the shipyard, Fiona and Robin knew every inch of her—but maybe they'd overlooked some mistake or flaw. If they had, the ocean would certainly find it.

As the hours passed in the hazardous seas, they noticed little things going wrong. Chafe constricted *MonArk*'s sails. Patches of engine oil were in the boat where they shouldn't have been. Both were relatively minor fixes on land, but now was not the time to investigate further. The little things rankled nonetheless, picking away at their confidence. Amid the stress and excitement of starting a journey across the Pacific, Fiona was exhausted. She hadn't figured out how to sleep well at sea, and the last four sleepless nights had worn her down. Along with the fatigue, she was feeling seasick, dizzy, and just plain scared. The ocean cracked big loud waves against the boat, spraying water into the cockpit, where Robin and Fiona sat holding hands. They both wore bright red survival suits. Harnesses clipped them to their boat. "The thought that a single mistake, like an unclipped harness, could wash either of us into watery black oblivion was terrifying," Fiona wrote later. She concentrated her thoughts on the sound of another wave crackling behind the stern as she surfed the boat down another towering wave. Her head swam with viscous thoughts, and her stomach churned with waves of nausea. Feeling slow and groggy, she faced down the most challenging conditions she had ever seen at sea and held the boat steady through the long night.

Joseph Conrad wrote that a love of the sea is actually a love of ships. I love the sea, but I don't love ships, so I think Conrad's point is actually that we fall in love with the ocean through an activity.

Sunbathers worship hot sand and the perfect amount of wind. Fierce coral communities duking it out between the polyps draw in divers. Paddleboarders love to skim the surface, horizon at eye level, ripples growing into waves. Fishers see the ocean through their prey, paying attention to river mouths, light sources, and the shape of the seafloor. For me sailing was the first way I loved the ocean. In a way, it didn't feel like a choice.

During the summer, my family fled the big city of Toronto for the Atlantic coast, where we spent hot, humid weeks in Cape Cod and the Maritimes. Along the way, my mother doled out a treasure trove of family sailing stories. Her grandmother (Nana to us) grew up in the booming port town of Halifax during the heyday of commercial steamships. Nana's father worked as Halifax's harbormaster, overseeing the ships making their first stop in North America and then continuing on to Boston or New York, or getting last-stop provisions before setting out for Europe.

Until her death at 101, Nana told stories about sailing the East Coast with her father, learning to navigate the rocky shoals that he called the hens and chicks and beating the rich kids in their fancy boats during regattas. In December 1917, when she was eighteen years old, two ships, one carrying munitions bound for European trenches, collided in the Halifax harbor. Nana vividly recalled how the windows of her home exploded that day. Just under two thousand people died instantly, and much of the city's north end was flattened. The city grieved and struggled to move forward in the middle of a cold snap right before Christmas. At the trial that followed, her father testified about the measure he had instituted of unloading explosives at an island outside the Halifax harbor. But the British army commander had taken over his position and stopped a practice that might have prevented catastrophe.

Nana's father came of age in an earlier time, just as faster coal-powered ships were eclipsing commercial sailboats. Offshore sailing was born at this juncture, in the 1890s, when a merchant marine captain found himself broke and out of a job. Joshua Slocum, originally from Nova Scotia ("It is nothing against the master mariner if the birthplace mentioned on his certificate be Nova Scotia," Slocum once wrote), rebuilt a rotting thirty-six-foot boat in a Boston shipyard and pitched the first-ever solo sailing circumnavigation to newspapers. He worked his way around the ocean, filing dispatches from exotic locations and igniting the public's imagination about finding oneself and adventure at sea. His accounts, related with manly understatement, captured a nascent modern era: here was rugged individualism in the flesh, a man taming the elements, traversing new terrain as he staged the second act of his life. On Slocum's return to land, he compiled his travels into the much-beloved and best-selling memoir *Sailing Alone Around the World.*

During the violent first half of the 20th century, offshore sailing struggled to take off as a middle-class sport. People who sailed off into the open ocean were rich or eccentric—and it was better if they were both. In the 1920s, Wimbledon tennis player Alain Gerbault sailed across the Atlantic with little experience or equipment. Twenty years later, Vito Dumas, an Argentine sailor, sailed through the turbulent Southern Ocean with newspapers stuffed in his clothing to keep warm. The public lionized these men as gods and champions. Sailing across the ocean, it seemed, was not meant for mere mortals. Their stories were canonized in the literature that I was falling in love with right around the time I started to write seriously.

For years, I had gorged on stories about the sea from Rachel Carson, Jacques Cousteau, Herman Melville, and Jack London. But it was the sailing writers I loved best: Joseph Conrad, Joshua Slocum, Bernard Moitessier, and the Canadian Kevin Patterson. Maybe sailors on long ocean passages had more time to write and reflect. Maybe something meditative about the ocean waves gave that writing a certain lyrical quality. Either way I loved their sense of adventure, the simultaneous journey toward self-realization and the ends of the Earth, the way they built a mystery about what lay beyond the horizon for those brave or crazy enough to sail there. I was hooked.

I read, I wrote, I started to sail, and all of this turned into dreaming of crossing the ocean. The inside flaps of the sailing books I read often began with a map of the ocean, the journey's route pencilled across the water. As I grew more versed in the literature, I began to see that line as a table of contents: an outline of the story arc and a hint at the stakes along the way. If the route ran along the East Coast of America, from Chesapeake Bay down the Gulf Stream toward the Caribbean, I was in for a vacation book. A visit to the sluggish Sargasso Sea would turn up more meditative, introspective writing in the "horse latitudes" named after Spanish conquistadors who ditched their horses to hurry their sails. If the line tracked through the Roaring Forties in the Southern Ocean, I was in for an adrenaline-jacked adventure: high stakes, high winds, maybe penguins. If that route went by Cape Horn, at the tip of South America, where two prevailing winds rush down the west and east side of the continent and form offshore sailing's Mount Everest, then I was definitely in for a record-setting adventure book. The sailor Miles Smeeton's book title even name-checked George Mallory's flippant explanation for summiting Everest: *Because the Horn Is There.*

The ocean can't hold a footprint or a flag, so that line across the map and the writing about the journey always seemed like a proclamation to me, the only way to testify to one's crossing. Like a bathroom scrawl: *I was here.* After years of mental preparation, of mundane collecting of facts about ocean routes and timing of journeys, I wanted to set out. And yet I was scared of sailing out of sight of land. "Adventure means risking something; and it is only when we are doing that that we know really what a splendid thing life is," wrote the early solo sailor Alain Gerbault. That was probably true, but I related more to David Foster Wallace's "intuition of the sea as primordial *nada*, bottomless, depths inhabited by cackling tooth-studded things rising toward you at the rate a feather falls." This "marrow-level dread of the oceanic" made it impossible for me to take any steps toward planning, toward leaving, toward one day "doing the dream." And then I ended up at a training seminar for people like Fiona and Robin, who were putting the dream into action.

❧

In a community center basement in Vancouver, the Bluewater Cruising Association offers a range of programs to support people planning a trip by sea. The Vancouver chapter has around nine hundred members, or between four hundred and five hundred boats, and membership is divided into three categories: Doners, Doers, and Dreamers. Sailors who have logged more than 999 nautical miles outside of territorial waters are Doners and they receive a small white pin with the word stamped on it when they return. The Doers are never seen at community events because they are out somewhere on the ocean, doing the dream. Then there are the Dreamers, who are planning to leave sometime in the near future but have not necessarily set a specific departure date.

An outsider can find it hard to tell the difference between a Dreamer and a Doner. "The people who have done it, they usually look very neat and tidy," Gillian West, one of Bluewater's older and most experienced sailors, observed. "Then you see somebody, and they come in wearing a watch cap and boots and a pipe and the beard, the whole works, and they've never been offshore." She laughed. Offshore sailors who have gone through customs and immigration know how important appearance is to border officers, who are suspicious of any signs of untidiness, she explained. "I remember meeting offshore sailors for the first time and thinking, gee, they seem kind of ordinary," another Doner said. She was a disarmingly warm occupational therapist with soft blue eyes.

She and her husband had sailed twenty-five thousand nautical miles across the South Pacific for three years. They weren't burly high-sea sailors or wealthy blue bloods with vermouth in their veins. They looked like any couple in their midfifties with two grown children.

I arrived ten minutes before the seminar began, and all forty students were there, seated at three long tables, notepads and pens out. After a full day's work, everyone here chose to sit through a three-hour-long lecture on marine electronics.

I had recently started attending talks about sailing across the ocean. At one, the speaker spoke about an impulse trip he made across the Atlantic, but his details were too vague. Just fly to Trinidad. Walk the port until you meet two pot-smoking Frenchmen. Learn the ropes as you coast through the islands and make your way to open ocean. "Sailing is there for the taking," he concluded with an anyone-can-do-it shrug. At another lecture, a retired professor narrated a slide show of his circumnavigation of Japan: picture after picture of sea meeting sky, mountaintop hot springs, jungles of three-thousand-year-old cedars, and buffets of fermented food and freshly caught fish. Because he sailed into the island country, there was no required landing at a major airport, no mention of any Japanese city a Western tourist might recognize. Via the ocean, he went directly into a hidden world of remote towns and villages emptied by urbanization.

Tonight's training session was of a different order than these splashy storytelling nights. This was Fleet: a subset of training seminars from Bluewater intended not for armchair travellers like me but for people with tangible plans to leave in the coming year or two. Fleet focuses on hard skills like navigation, weather forecasting, rebuilding diesel engines, customs procedures, and the effect of a changing climate and warming ocean on storm patterns. Some details are constantly in flux. Marine electronics improve on a yearly, sometimes monthly, basis. Other skills are constant, like how to pick up a mooring buoy at sea while cruising into harbor under sail or heaving-to (stopping) in a storm.

Since the 1970s, the organization has transformed thousands of seemingly ordinary people into ocean-crossing adventurers. It was here that I first met Fiona. At twenty-seven, she was the youngest person in the room. She had been dreaming of sailing the ocean since she was a little girl she said to me in the hallway during a break, as people nibbled Rice Krispie squares. At university, she wrote essays about making a trip by sea, and although she pursued a career in strategic finance, she included the dream in her five-year plan whenever someone asked. But how did actually deciding to do the dream happen? I asked her.

"It's a real psychological shift when you decide, okay, we're going," Fiona said. Over Christmas 2014, she and Robin announced the trip to their families, like one would an engagement or a pregnancy. Three months earlier, she'd quit her corporate job to pursue children's writing full-time and prepare the boat. In April 2015, Robin would join her, taking a leave of absence from his position as a building science engineer. He would start repairing and provisioning the boat full-time with Fiona.

In the best-selling memoir *Wild*, Cheryl Strayed describes a constant recommitment to her goal of hiking the Pacific Crest Trail: "There was the first, flip decision to do it, followed by the second, more serious decision to *actually* do it, and then the long third beginning, composed of weeks of shopping and packing and preparing to do it. There was the quitting my job as a waitress and finalizing my divorce and selling almost everything I owned. . . ."

Each step of turning the dream into reality calls for yet another level of sacrifice or investment. At some point, you have to stop questioning whether or not you're really going to do the dream, and just do it. Six months before their departure date in the fall, this was the stage Fiona and Robin were navigating. When Robin told people at work that he was leaving to sail the South Pacific, they responded with shock and surprise, more than he received from his own family. "Can you really do that? Like, is that possible?" one flabbergasted coworker asked him. A guy in marketing said, "I would never do something like that. Way too risky." But, Robin noted, "These are pretty smart people. He thinks it's crazy for *him*. He doesn't think it's crazy for *us*."

Robin and Fiona both come from adventurous families. A circumnavigation is not out of place. Fiona's uncle has circumnavigated twice. Robin spent a part of his childhood travelling in a van with his parents through Canada's Yukon Territory, living in a tent or squatting. Their upbringings predisposed them to sailing across seas, but Robin and Fiona were still working through the many physical and mental challenges of doing a journey entirely on their own.

As one of Bluewater's youngest members, Fiona saw offshore sailing a little differently. Nearly every other boat in Fleet had either a single retired male sailor or a retired couple on board. These Dreamers have a lifetime's worth of experience and savings behind them to make the dream finally happen. Fiona wanted something different: to make a life out of their trip and shed the nine-to-five lifestyle once and for all. This put even more pressure on getting the trip right. It wasn't just a fun vacation that she could stop at any time. It was a reinvention of a routine that most people in the world live by. In one push from the dock, they would cast off the whole rhythm of land: bimonthly paychecks, monthly bills, weekly meal planning, Friday drinks, Monday morning emails, peak traffic times—everything that is both comforting and claustrophobic about life on land. At Bluewater, they were learning to be reliable at sea and hoping that they could become financially independent there, too.

"The reason money is not such a concern for us is that it's very inexpensive to live when you're out there cruising," Fiona explained. "Most people say a thousand bucks a month is pretty good, like you're comfortable, for two people, and that's nothing. Even though

I'm only pulling in a hundred bucks a month from my website or blogging, it's actually meaningful, as opposed to a drop in the bucket."

Depending on where you cruise, costs can run from $30,000 to $70,000 every year. In many of the world's most expensive cities today, that's the bare minimum needed to live. Between them, Fiona and Robin had saved for the first two years of their trip, but a large chunk of that would go toward boat repairs. Another advantage on their side was that they were frugal, good at budgeting, and had complementary skills: Robin's engineering background, and Fiona's financial acumen and family history of sailing.

After work and on weekends, they were consumed with the boat, the journey, and everything it asked of them. They repaired the worn-away plumbing or went to shipyards to shop for spare parts, or they attended training seminars like the ones at Bluewater. They amassed dozens of scribbled to-do lists and prospective budgets and Excel spreadsheets with firm deadlines for what tasks had to get done and when. Each day began with a new to-do list and they worked twelve-hour days scrambling to get it all done in time.

Because everyone has to start as a Dreamer, Bluewater classed them as that. Fiona has been a Dreamer her whole life, but she and Robin didn't like the label. A true Dreamer is someone like me, who attends talks and holds vague plans of going offshore someday. They preferred Leaver, which is not an official designation at Bluewater, because at least that term recognized their serious intention to cross the ocean. They felt other members understood what they meant when they said it. A Leaver was someone who'd made the commitment to leave within the next two years. The term bridged the division between doing and dreaming that runs through the culture. Hal Roth's classic offshore sailing manual *How to Sail Around the World* devotes an entire chapter to the subject. All the endless spreadsheets and shopping have an aspirational element. Planning is a form of dreaming. "The planning stage of a cruise is often just as enjoyable as the voyage itself," wrote Jimmy Cornell, the British yachtsman and author of *World Cruising Routes*, the most respected authority on offshore sailing routes. "We have a saying about that," a veteran sailor at Bluewater told me. "You prepare and prepare and prepare. We say, 'Don't forget to go!'"

"A five-year timeline you hear pretty often," Fiona said. "Which I think is pretty interesting, because it's just long enough that you don't need to be doing anything about it right now."

"You don't have to have a boat with a five-year timeline," Robin said.

"When you get to the three-year mark, you have to start making decisions that mean money, time, and getting ready," explained Fiona.

"But people do it in compressed timelines," Robin added. He sometimes worried that when they told people they were going in one year, "They think we're not going to go, because our timeline is so compressed."

❧

Offshore sailing is safer today than at any other time in the history of the sport. After World War II, cheaper plastic and fibreglass boats made their way onto the market, opening the sport to middle-class hobbyists. Marine and hardware stores sold build-your-own boat plans that people constructed in their backyards. That was how Gillian West, one of Bluewater's most august members, was drawn to offshore sailing. In the 1970s, her husband brought home a DIY boat plan, and together they joined this early generation of hobby sailors sailing offshore.

Today's offshore sailors can be in the middle of the ocean streaming a yoga video or uploading a blog post. The constant connectivity mimics life on land, and the escape that the ocean offered feels distant, a thing of the past. Of course more advanced electronics also make offshore sailing safer than before. They're a lifeline at sea. But just a few decades ago, sailors were still navigating with the stars, using heavy sextants just as seamen did in the 18th century. There were no fridges on board. Most boats didn't have toilets and used only rudimentary electronics, like a trailing log that ran off the back of the stern, counting nautical miles as they sailed. But given enough time and dedication, the average person could make the leap from armchair sailor to circumnavigating adventurer. In her seventies, long after her husband had lost interest in sailing across oceans, Gillian West circumnavigated the world, hiring crew to trade night watches with her along the way.

"Now the electronics are just amazing," Gillian said. Modern boats today come equipped with satellite internet and phone, as well as the AIS ship identification system that pings a ship's location to a satellite and back down to Earth. "You'll get a blip, and you'll put your cursor on it. It'll tell what the ship is, the direction it's going, its speed, how close it's going to be to you, as well as cargo, coming from and going to—probably more than you'd ever want to know."

On marinetraffic.com you can watch boats move across the ocean in real time using the AIS ship identification system. Thousands of little dots: pink for sailing boats, red for cargo, orange for fishing, green for government, on and on. Watching the dots move like a flock of birds across the screen, following shipping routes and avoiding low-pressure systems, I felt some of my romanticism about sailing die. Luxury boats today are also kitted out with air conditioners, freezers, and satellite TV. Cruising off into the horizon is no longer what it was back in Joshua Slocum's day. When the founding father of offshore sailing left for sea, his only communication was mailed dispatches from exotic ports back to the United States that took weeks to arrive.

For some, Slocum's journey was not extreme or isolated enough. In 1968, the British newspaper the *Sunday Times* sponsored the first-ever sailing circumnavigation race: the *Sunday Times* Golden Globe. This circumnavigation was solo and nonstop, meaning no stops in port, no supplies taken on board, and no repairs unless the sailors performed them at sea. The Golden Globe gave birth to unassisted sailing, the most extreme no-holds-barred version of the sport, and forced isolation on the nine participating sailors. One, Bernard Moitessier, had to slingshot messages to a passing freighter when he communicated with the outside world.

Another participant, Donald Crowhurst, went to sea fantastically underprepared. Fourteen days in, he had to face reality in the form of a long list of boat repairs, some serious, some minor, but all pointing to the logical conclusion that his boat was unready for rough sail in the Southern Ocean. Instead of backing out, he circled the Sargasso Sea, waiting for his competitors to round the globe, hoping to slip in with the pack as they approached the finish line. Long before that day came, he cracked under the pressure of pulling off such an elaborate fraud. His boat was found adrift at sea; the notes left behind showed a painful slide into madness and then suicide. Six other competitors had already dropped out of the race. The front-runner, the mystical Frenchman Bernard Moitessier, came to resent the race's growing commercialization of offshore sailing. After rounding Cape Horn, he snubbed the race's finish line in England and continued sailing another three months and 180 degrees around the world to Tahiti. Nigel Tetley's boat sank 1,200 nautical miles from the finish line, and only Robin Knox-Johnston successfully completed the race.

Crowhurst's fate shocked the readers of the *Sunday Times*. To insiders, however, many warning signs portended disaster. Some were legitimate concerns, like Crowhurst's lack of sailing experience, his haphazard navigation skills, and the disappointing performance of his trimaran. Others were bad omens and superstitions, like the burn on Crowhurst's hand that destroyed his lifeline and worried his wife before he sailed off.

Crowhurst's ill-fated journey is part of sailing lore, but his early setbacks and disappointments are not that different from the beginnings of happier sailing stories. "Almost every account of a successful adventure contains in its early chapters instances of over-optimism, confusion, and pushing salesmanship," note Ron Hall and Nicholas Tomalin in *The Strange Last Voyage of Donald Crowhurst.* "Once the happy ending is reached these early setbacks only look like evidence of determination and integrity."

The Vendee Globe, as the Golden Globe is known today, looks very different from the 1968 race. Each race has become a proving ground for new technology. The introduction of emergency location devices, like EPIRB, has decreased deaths dramatically. But too much technology can deceive sailors into thinking they're more connected than they truly are. Some speculate that all the new gadgets push sailors to extremes and encourage people to take more risks. Robin Knox-Johnston, the winner of the 1968 race, has argued that taking away a sailor's EPIRB might encourage caution because the sailor couldn't take a risk, then call for help.

The Fleet seminars are meant to teach self-reliance at sea, to prepare sailors for what to do when they can't call for help. They might have to jerry-rig a solution from tools and gear on board and then double- and triple-check those repairs. This means acquiring lots of big and little skills, but also a new frame of mind. Life on land in an urban, developed city has a safety net that is so subtly pervasive we barely notice it anymore. Guardrails and yellow lines keep us back from the edge; smoke alarms and carbon monoxide detectors monitor the air, and if all else fails, we can dial 911. Going offshore means abandoning that safety net—something very few of us experience in our lives.

Out at sea, Fiona and Robin would have to rely on each other. When things went well for them, the ocean meant total independence and freedom. When things did not, it could mean danger and even death. Bluewater's take-home message on marine electronics was that technology is no substitute for seamanship. The instructor gave one more piece of advice: install backup systems for backup systems, he said. Something will go wrong at sea.

❧

Robin and Fiona's biggest fear of sailing across the ocean was their boat. They lived on it in one of the last marinas in Vancouver that still allows people to live on board. They managed to find a space by leasing a berth from a sailor who was taking care of her sick mother on land.

I was buzzed through the marina's electronic gate and ran into Gillian West, who was walking the docks in electric pink sneakers. With so few marinas left where they can live on board, this group of offshore sailors forms a tight-knit community of liveaboards, most of whom attend the Bluewater events and lectures. Newbies sail blithely past the marina with the wrong lights on port, the wrong lights on starboard, while the neighbors on the docks gossip about the mistakes they see boaters make on the water. I told Gillian that I was visiting Fiona and Robin, who were busy preparing their boat for their South Pacific sail. "People always say it's easy to sail around the world," she quipped. "The hard part is taking the boat."

Just as I found their berth number, Fiona poked her head up from *MonArk*'s hatch. She welcomed me aboard, and we ducked down into the galley. Like most sailboats, it was dark down below but cozy and furnished entirely in gleaming dark wood. She took a seat at the galley table, where she was drinking a mug of steaming tea and waiting for Robin to get home.

"We worry about her," Fiona said, looking around, "but she's made the trip before." *MonArk*'s past owners sailed her to Australia and then on to the west coast of North America.

The hatch above slid open, and rain from outside pelted down. A tall figure in a rain jacket came down the ladder and removed his hood. Robin was home. He peeled off his wet jacket and joined us in the galley.

With only a few months before they left, Robin and Fiona were still flip-flopping on whether to sail *MonArk* or find another boat. Things kept going wrong that shook their confidence in her. Robin poured grease down the sink and the hose underneath clogged. When he went to fix it, the hose disintegrated in his hands. This led to the discovery that the entire drain assembly underneath the kitchen galley had corroded. It seemed like everything they touched on *MonArk* seemed to fall apart.

"There was this domino effect of things going wrong, and it got me so worried," Robin said. "I went to bed that night thinking, that was from pouring grease down the sink, and that's a non-threatening issue to us—what else is there that we don't know about?"

This domino effect also goes by the name cascading failure. It's a common phenomenon in electrical grids: one node fails, overloading a nearby node and causing it to fail, and so on. But it's also a good metaphor for how a seemingly simple and benign problem on a sailboat, like a clogged sink, could morph into a more serious one at sea. The 2013 film *All Is Lost* is a perfect portrayal of cascading failure. Robert Redford plays an offshore sailor whose boat is damaged by a drifting shipping container in the middle of the Indian Ocean. Robin and Fiona watched the film together, as Redford's character tries again and again to save himself and his boat. "We came away from it thinking he seemed really prepared, he did a lot of smart stuff," remembered Fiona. But then they went to a training seminar at Bluewater and asked the more experienced sailors what they had thought of the film. That didn't go well. In sailing circles, *All Is Lost* is almost universally reviled. Sailing blogs and magazines are rife with *All Is Lost* reviews that pick apart each decision the sailor makes, from the laughable repair job he does on his boat to his poor etiquette leaving the hatch open each time he goes on deck. Worst of all, he ditches his sinking boat when he could have made many more, and better, attempts to save it.

"Everyone said, 'Oh what an idiot. You would never leave your boat and get in a life raft,'" Fiona laughed.

"We were like, yeah, totally," Robin says.

The odds of a shipping container slipping off one of the thousands of freighters somewhere in the ocean and colliding with an offshore sailor's boat were so remote, one Bluewater sailor insisted, that if you worried about those things, you'd never leave land.

This is the advantage Doners have over Dreamers. Dreamers see an ocean of potential dangers, while Doners know from experience what might go wrong and what has so little chance of going wrong that it's not worth losing sleep over. Redford's character's choices don't make sense to offshore sailors who had actually been out there, but that distinction was so subtle that it wasn't obvious to Fiona and Robin. The only way to *really* know all those nuances is to go try it yourself.

In the weeks leading up to departure, a Dreamer's physical and mental preparations begin to collide with reality. Is the boat safe enough? Are the weather conditions right? Even superstitions, something most sailors won't admit to believing in, gain real-life currency right before the crucial departure date.

"I would consider myself super non-superstitious," Robin said. "I would happily walk under a ladder, but not right before we leave. I'd smash a mirror, and I wouldn't even care, *unless* it's right before we leave. No black cats. I think superstitions are a way to gain control of your life when you're afraid of something."

Fiona compared their departure date to a wedding day, which also happens to be an event people build elaborate superstitions and rituals around as a way to influence the future. Eventually the day arrives, and the Dreamer has to decide whether to go and meet reality or stay and keep on dreaming. In early fall, before hurricane season arrived, Fiona and Robin would make their final decision, too. Maybe the boat wouldn't be perfect; maybe they would set sail on Friday the thirteenth. Either you leave, or you keep dreaming.

On a warm summer evening a few weeks before Fiona and Robin's departure, Bluewater held a farewell dinner for that year's Fleet. In the marina where they lived, high energy coursed through the clubhouse as the last of a summer day faded through the window. Forty sailors had spent the last year learning and training together in Fleet. Now the time had finally come for them to sail off into the horizon. Some had already left. Every fall, wild weather hits the Oregon and Washington coasts to the south. By October, the perfect weather window to sail south for Mexico has closed. If sailors see a stretch of fair weather days before that time, they take it.

The small room filled with Dreamers and Doners, passing around bags of chips, Tupperware containers of pasta salad, and homemade desserts. Also passed around were final good-byes and hugs and last-minute trip suggestions, like a great grocery store in Point Reyes, California, or where to find cheap fuel once they reached Mexico. The dream was finally happening.

A group of Doners had organized the party, and they seemed almost as excited as the Dreamers. One Doner who had become a mentor to the Dreamers talked about what sailing across the Pacific had given her: a sense that she could do something she thought was impossible. "That doesn't happen to me a lot at home, but it did out there," she said. "That passage coming home, it was eight thousand miles. For me, that was a huge accomplishment. It even helps me now when I'm doing something that I don't really want to do, or I'm trying to push myself, and I think back, well, if I can cross oceans, then I can do this."

A few weeks earlier, Fiona and Robin had hauled their boat out of the water for the final round of repairs at a nearby dockyard.

Since then, they had lived in the dry-docked *MonArk* on land, but they already looked the part of vagabond sailors. Their clothes were rumpled and paint-smudged. Robin's brown hair and beard had grown in long and shaggy. Fiona's face had taken on a sun-kissed pink. They looked distracted, but happy.

Every day that ticked past moved them a little bit closer to that fateful date, when all those spreadsheets and to-do lists had them set to leave. Unfortunately, *MonArk* was in much worse shape than they thought. The balsa deck was a lot more damaged than expected. They ripped out and replaced most of it. New rigging was another expensive investment, as were all the additional electronics. The repairs became so costly that they stopped talking in thousands of dollars and simply referred to them as "boat units." They placed *MonArk* in the ocean after the extended haul-out and discovered water leaking in. Another haul-out, another three weeks onshore, a few more boat units.

That night, Fiona and Robin were the first to leave the party. They still had laundry to do, last-minute possessions to cram on board, and many more errands to run before they finally set sail. The boat still might surprise them with some new and expensive repair. And yet Fiona remembers those frenzied last days as some of her fondest. They were riding the high of leaving while feeling the crunch of money and time. In short, they were living.

Watching the Dreamers make plans to meet up in the Sea of Cortez and take their leave, I saw departure normalize in front of my eyes. Some people in that room were probably more prepared than others, and those underprepared Dreamers would make more mistakes along the way. That wasn't any reason to not take on the trip, I realized. It was the process that mattered. All that planning and preparing had become less terrifying for me, less filled with life-or-death consequences.

Sadly, this meant sailing across the ocean seemed less grand in my mind. Laying bare all the menial and repetitive tasks behind it made it seem accessible and boring. Gaining the reality meant losing the romance. When I stood on the shoreline and looked out at the sea, my family history of sailing, my personal love of sailing, and sailing stories all came together, and I felt . . . better. A day out on the water washed away tension bottled up inside me and relaxed knots of anxiety twisted in my stomach. I looked at the ocean, contemplated my tiny space on a planet made of water, and I saw infinity.

After two miserable days battling the towering waves they encountered off the coast of Oregon, Fiona, Robin, and *MonArk* will survive. As they sailed away from Cape Blanco, the twenty-five-knot winds will drop, in an hour, into a calm three knots. Absolutely exhausted from the ordeal, they'll head for land, motoring quietly through the fog, and make landfall in the small town of Eureka, California, just over the state line. It was 300 kilometers north of San Francisco, their intended first stop, but it would do. "We pulled into Eureka, California, with our engine spewing oil, a battered sail, and bruised egos," Fiona wrote in an article about the trip in *Canadian Geographic*. "Humbled, we went about our repairs, trying to ignore the cooling air, an ominous harbinger of the Arctic lows that would soon begin their relentless march down the coast." Apart from a dicey stretch of wind and water known as the Dangerous Middle just outside Samoa, this will be the worst weather they encounter in nearly two years at sea. Other far bigger surprises along the way await them. During their first month in Mexico, Robin's father will pass away suddenly. Not long after, they'll take some time off from the journey to work and save more cash, all the unexpected repairs and expenses having blown through their budget. They'll leave *MonArk* at a marina in Mexico and return to British Columbia, where they'll spend six months with Robin's family in a northern town, working odd jobs like cleaning houses, building decks, and weed-whacking cemeteries. A hurricane in Mexico will rip through their marina, toppling a line of boats, including *MonArk*. They'll fly down to repair and ready her for their biggest trial yet: crossing the South Pacific from Mexico to the Marquesas. Before they leave, they'll marry near the town of La Cruz de Haunacaxtle—a popular place for offshore sailors to start the crossing. Late in 2017, they'll arrive in Sydney, Australia.

As I said good-bye to them, I thought about how their dream would not be what they dreamed it would be. But they would do it anyway, despite the dangers and doubts. The Dreamer would become the Doer, and the Doer would become the Doner. And then the Doner would find a new dream.

THREE

CROSSING THE MEDITERRANEAN

No pain can shake a man as badly as the sea, however strong he once was.

—Homer, *The Odyssey*

In 2015, Hassan Basheer stood on a dark beach in Turkey and watched his smuggler pump air into an inflatable white raft. It was then that the twenty-six-year-old Syrian realized he was truly going to cross the Mediterranean on a crude boat. Two months from then, a toddler's body would wash up on a beach not far from where Hassan stood. The Turkish photojournalist Nilüfer Demir would capture on film the moment a policeman scooped the toddler's body off the wet sand. The photograph would go viral, circulating the internet millions of times over, as would the toddler's story. Alan Kurdi, a three-year-old Syrian-Kurdish refugee, would drown when his overcrowded sixteen-foot dinghy overturns in the Mediterranean. Kurdi would become a symbol for the hundreds of thousands floundering at Europe's border, but for the moment, as Hassan stood on the beach, the world remained oblivious to secret journeys like Hassan's and the boats crossing the dozen or so kilometers of ocean that separate Turkey from Greece.

Hassan had borrowed money from friends to pay for the journey. He gave three thousand euros to a Syrian smuggler, a middleman who made money off recruiting refugees from his own country while he tried to save enough money to cross himself. The man assured Hassan that the trip was perfectly safe. No more than thirty-five people on board a boat built for fifty, he promised. There would be life jackets and the journey would be quick. "We do this every day," the smuggler shrugged casually.

After they paid, Hassan and his friends waited for the smugger to call. They spent three tense, tedious days in the resort town of Izmir, trying not to overspend, bumming around a Syrian restaurant called Sindybad (Arabic for Sinbad, the sailor) with all the other refugees waiting to cross. He never stopped turning his choice over and over in his head. Could he really get on the boat? Was he paying his friends' money to drown in the Mediterranean? Could he back out now and still get the money back? He found the journey impossible to contemplate. He couldn't swim. He'd never seen the ocean or set foot on a boat before he arrived in Turkey. The sea frightened him.

On the beach, the smugglers herded people on board the three rafts by the aliases of their recruiter. "Abu Ahmad!" they called out. "Abu Omar!" ("Father of Ahmad" or "Father of Omar.") During the sorting, Hassan considered running off into the darkness, but he had no idea where he was or how to find his way up the valley, cloaked in night, and back to Izmir. The smugglers were rumored to beat anyone who refused to board. He hesitated a moment longer, and then he and his friends were shoved onto a raft together.

A year later and across the sea, where the Mediterranean washes up on the bleached shores of Libya, another young man also waited. Mohammed Botwe was only seventeen, but his voice and bearing belonged to a much older man. The tall, deep-voiced Ghanian had spent the last month in a squalid camp outside Tripoli. This place was called a *mazraa* (farm). Run by smugglers, these way stations treat the refugees who pass through them like animals. A *mazraa* can be an abandoned farmhouse, an unfinished villa, an old warehouse, pretty much any structure, as long as it is close to the coast and can hide hundreds of people from the Libyan authorities, who may or may not be in on the deal. In Mohammed's *mazraa*, he slept in a long, open room, crammed in with hundreds of other migrants, most of them from Senegal and Gambia, with only a few inches between them. The floor was covered in dirt, garbage, and human waste. The *mazraa* offered no drinkable water, no showers, nowhere to wash up. People drank the seawater. In the evening, the smugglers distributed a meager meal—a bowl of thin gruel shared by dozens—that barely sustained the migrants through the following day. All the better for the smugglers' profits to keep their cargo light and cram ever more migrants onto each boat. No one knew when the boats would come.

As the days passed, Mohammed felt desperation in the *mazraa* growing. Stories circulated of people waiting months before they got a place on a raft to Italy. "We will have to fight to get a spot," he told the only friend he'd made, another Ghanian. One morning, two black boats arrived at the camp. It was an early morning in April, maybe around seven, and Mohammed wondered why they weren't leaving at night, when the authorities wouldn't see them. The hundreds of people who had been waiting at the *mazraa* rushed to the beach.

The dinghies were less seafaring vessels than floating platforms. They measured around a dozen meters long with an inflatable pontoon-like hull on each side. Anything could pierce the soft inflatable sides of the boat, and the smugglers had told the migrants to remove anything pointy or sharp before climbing on board. They left chains, earrings, buttons, cellphones, and shoes scattered on the shore.

Everyone started pushing to board the two boats. Mohammed lost track of his friend, but he figured the man would make it on the second boat. He waded out to the first boat, the water to his waist, and without waiting for permission to board, he swung one long leg over the gunwale. His foot slid forward inside the boat, and his pants ripped open at the crotch. He felt his bare ass swing out into the air. The women with covered heads averted their eyes. Mohammed didn't care. He pulled himself up. He was on the boat; he was going to Europe.

Hassan Basheer and Mohammed Botwe were of different races, nationalities, and backgrounds. Different circumstances brought them to the Mediterranean. But they were like many people who risked their lives crossing the Mediterranean in recent years: young, male, and from predominantly Muslim countries. Their crossings were neither cheap nor safe, but nothing better waited for them back home. If Hassan or Mohammed fell overboard, or if their boats capsized or sank, they would become a number, one among thousands lost at sea.

The Mediterranean's extremely literal English name says exactly what it is: a sea "between the lands." Just over twenty nations share its 46,000 kilometers of coastline. The close proximity between countries of such stark inequalities is perhaps the simplest explanation for why people cross. The same azure water that washes up on the French Riviera and the party island of Ibiza also arrives on the Assad regime's port town of Latakia, Syria, and the formerly ISIS-controlled city of Derna in Libya. The UN's International Organization for Migration estimates that 15,000 people have died migrating across the Mediterranean since 2013. Because smugglers arrange most of these trips, no one knows exactly how many boats set out or how many people are on them. Criminology professors Leanne Weber and Sharon Pickering write that for every body found, two more lie beneath the waves, according to expert estimates.

Water's slippery materiality makes it a difficult place to police and an ideal place to cross borders. The term *boat people* has applied, sometimes dismissively, to many nationalities and crises over the last half century. The Vietnamese boat people, numbering around 800,000, fled across the South China Sea after the fall of Saigon to the Communist government in 1975, money and jewelry sewn into their clothing. The Cuban boat people have arrived on rickety rafts or boats in Florida ever since Fidel Castro's takeover in 1959, with the biggest surge in 1980 during the Mariel boatlift, when around 125,000 Cubans came to the United States by boat. A small group of the Marielitos, as they came to be called, had a criminal past or a history of mental health issues. This created the narrative that Castro purposefully dumped undesirables on American shores and overwhelmed Florida with Cuban crime—a myth that Castro encouraged and was later popularized in the 1983 Al Pacino film *Scarface*. The Marielitos are still raised as a spectre in anti-immigration rhetoric by American politicians today. The largest human migration in history also happened by boat, between 1815 and 1930, when fifty-six million Europeans crossed oceans to North America and Australia. This group, of course, has never been labelled *boat people*. More recently, Australia has instituted a zero-tolerance approach to asylum-seeking boats since 2001 when it introduced the Pacific Solution. This arrangement, and others like it by successive politicians, has pushed boat people arriving from across Southeast Asia and the Middle East into secretive offshore detention centers in Papua New Guinea and other nearby island nations. Elisa Kaltenbach is a PhD candidate at the University of Constance, in Germany, who studies trauma in refugees. "For us, it's not new," she observes. "I've been working in this context before this whole 'refugee crisis' came up." Her fingers marked air quotes around the phrase. "They've always been coming."

For centuries, the waters of the Mediterranean Sea have been a site of migration, but it used to move in a multitude of directions, not simply south to north. Throughout the 18th and 19th centuries, southern European peasants sought safety and opportunity in northern Africa, and northern Europeans did the same moving to Istanbul during the Ottoman Empire. But the surge in clandestine boat-crossings arranged by smugglers is a new and dangerous phase that started a few decades earlier. In the 1990s, during the civil war in Albania and the Kosovo conflict, things began to change. Refugees started to cross into Greece and Italy. In March 1997, the Italian Navy rammed the *Kater I Rades*, sinking it and drowning eighty-one Albanian refugees on board (only fifty-seven bodies were ever recovered). The Italians insisted the collision was accidental, but scholars cite this incident as a turning point toward a more militarized European coastline. Meanwhile, conflict along the eastern and southern shores of the Mediterranean has deepened, and more and more migrants would rather face hardship in the wealthy, stable countries on the other side of the sea.

For people in the developed world, trusting your life and future to a rickety raft is unthinkable. So are the experiences that lead up to that decision. When migration experts talk about why people leave a country, they break down the reasons into push and pull. Push factors drive people out: violence, persecution, poverty. Pull factors are what the destination offers: safety, freedom, a brighter economic future.

This combination of push and pull pulses across the Mediterranean, needs and wants, fears and desires. The Greek island of Samos sits a mere nautical mile away from Turkey, where thousands of Syrian, Afghan, and Iraqi refugees once tried to cross. On a clear day, they could stand on a Turkish beach there and see Greece, the prospect of a safer, better life in Europe tantalizingly close. The Greek islands of Chios, Lesbos, and Kos were also popular landing points, under a dozen kilometers from Turkey. Some refugees swam these distances, and their pictures were shared around Facebook like hometown heroes. The arrival of refugees on the white sand of Greek beaches has led to bizarre collisions: groups of bikini-clad tourists helping life jacket–clad refugees climb out of overcrowded rafts.

The majority of people rely on smugglers. In 2015, the industry grossed an estimated 5–6 billion euros in 2015. This was a record-breaking year in migrant crossings at all the EU's borders, but especially at the watery one between Turkey and Greece, where nearly a million people, Hassan among them, crossed the sea. The year before was also a record-breaker, but a modest one compared to the twenty-fold increase in 2015 between Greece and Turkey. In 2016, the pattern shifted: the route between Greece and Turkey dropped dramatically, partly due to an agreement in which the EU gave Turkey refugee aid in exchange for securing borders and accepting returned migrants. Meanwhile, the flow of mostly African migrants taking the longer and more dangerous route between Libya and Italy held steady and eventually surpassed ocean crossings between Greece and Turkey in 2016. Every year, the International Organization of Migration makes an educated guess at how many people go missing in the Mediterranean Sea. In 2016, the year Mohammed crossed the sea, over five thousand people died on the journey—the sea's deadliest toll to date.

With more deaths than any other migrant crossing in the world, including the US–Mexico border, the Mediterranean has a new but equally literal name in the press: a mass migrant graveyard.

Mohammed, like many Mediterranean migrants, had already made a long journey before he reached the coast. When he was ten or eleven years old, Mohammed realized he would never be safe in land-locked Kumasi, one of Ghana's largest cities with two million people. To cross one leg over the other or to hold hands with his partner or kiss him on the lips: these are dangerous acts in his country. The state punishes homosexual men with three-year-long prison sentences, while violence against them is tacitly condoned. The continent surrounding Ghana is not much better; over thirty African countries actively criminalize homosexuality.

In Kumasi, Mohammed had a live-in boyfriend. The neighbors noticed, but Mohammed only realized that when it was too late. One night a crowd broke into their home. Close combat followed, hand-to-fist fighting, knives slashing through the air. "True struggling," he called it. Mohammed somehow fought his way out of the house and onto the street. He escaped with a two-inch stab wound through his right shoulder. It healed on the death-defying migration he started that night. First he went south to Accra, where the smuggling route out of Ghana begins. Then he went through Burkino Faso and on to the dusty smuggling hub of Agadez in Niger, the last stop before the Sahara desert. This is a road riven with its own dangers: rapes, murders, kidnappings for ransom or human trafficking, people dying from thirst in the dunes. The destination is Libya, a country of prosperity, at least in comparison to the rest of Africa. If Libya doesn't work out, the country's coastline serves as the launching point where migrants crowd onto boats and rafts and hope to make it across to Italy.

But when Mohammed Botwe finally arrived in Libya after a long trek across Africa, the country was very different from the one he expected to find. In the months following the public execution of Libya's longtime dictator Muammar Gaddafi in 2011, tribes, gangs, and former revolutionary militias carved the country into competing fiefdoms. It was no longer a corrupt but wealthy dictatorship that ran on profits pumped from the oil fields in the Gulf of Sidra. When Mohammed arrived in late 2015, Libya was hurtling fast toward insolvency and chaos. His dark skin marked him as an outsider, an African, someone the authorities didn't care about. He was easy prey for kidnapping—or worse. "To see a gun in Libya is easy," Mohammed said.

He found a job at a car wash in Tripoli, where he worked and slept and was even paid occasionally if the boss was around. If Mohammed asked a customer for payment after finishing a car, the customer simply opened his jacket and revealed a gun or a knife inside. "You want your life or the money?" they would ask him, and he would flee. Mohammed didn't consider himself a slave in Libya, but he was certainly trapped.

One evening in early April 2016, Mohammed was scrubbing the last car of the day when the customer shouted that money was missing from his car's glove compartment. Mohammed's coworker, who was supposed to be helping him with the car, had also vanished. The customer and Mohammed's boss wouldn't listen to his explanations. They told Mohammed to deliver the thief or go to jail in his coworker's place. As a lowly migrant worker, Mohammed knew his story was worth nothing next to the word of two Libyan citizens. He stuffed all the cash he had saved down his boxer shorts and devised a new plan quickly. Until that moment, he never considered leaving Libya for Europe.

He led his boss and the customer to a house where African migrants hung out. He told them that he had spotted his coworker inside and that he would go in and flush the man out. "Wait here," he told them. Mohammed went inside and then escaped out a back door, running into the night. He hid until morning, when it would be easiest to pass through Tripoli's many military checkpoints. He didn't know what would happen to him if the police caught him, but he knew he needed to get out of Libya. Moving through the city, he finally reached a Ghanian neighborhood, where he could speak Hausa, his native language, and ask for help.

Human smugglers in Libya assume that Africans have less money and charge them accordingly. Mohammed used his savings to pay the roughly $650 for a spot on a boat to Italy.

After Mohammed paid his passage to a smuggler, he was stuffed into a car with eight other Africans headed for Europe. An elderly

Libyan woman sat in the passenger seat to ease the car's passage through checkpoints. Exhausted from the last days on the lam, he fell asleep under a coat on the drive. At one point, the driver started yelling in Arabic and woke him up. Mohammed couldn't understand the insults. Apparently he had broken something with the driver's seat while he was sleeping. He begged the man to take him to the *mazraa*, he begged like he was begging for his life. A seventeen-year-old Ghanian abandoned on a desolate road in Libya would not last long. Perhaps the driver took pity on him. Or perhaps he was more concerned with losing a lucrative job if he ditched a migrant in his charge. Either way, the man relented and drove him the rest of the way to the *mazraa*.

Mohammed was on the path to Europe, but he still didn't know if Europe would take him. That would have to be sorted out later. As the number of asylum-seekers crossing the Mediterranean grew, over a dozen EU countries drew up classification rules to process the newcomers. Ghana is considered a "safe country of origin" by many of the European Union countries he might want to end up in. Syrians, Eritreans, and Somalians have some of the highest rates of winning asylum in Germany, Britain, and France. But, as journalist Patrick Kingsley has pointed out, Libya turns people into refugees regardless of where they come from. In the chaos that is consuming the country, outsiders are the first to be exploited.

"Libya," Mohammed realized, "is a hell."

❧

Around 165 nautical miles (186 miles, or 300 kilometers) of water separate Libya from the European Union in the form of an Italian island called Lampedusa. Even for a yacht travelling at a speedy clip of fourteen knots (sixteen miles, or twenty-five kilometers, per hour), this trip would take half a day. For an overloaded dinghy with a small outboard motor ill suited to long ocean crossings, the journey can stretch into days—if the boat makes it. Smugglers don't usually inform their clientele that there's very little chance that an inflatable dinghy or rickety ship can get all the way from Tripoli to Lampedusa. Their best shot at survival is intercepting a ship travelling along the shipping lanes outside the twelve nautical miles of Libya's territorial waters. This reliance on third-party ships is built into the plan. During a fair-weather window, smugglers send out multiple boats at once each day. A passing ship has a better chance of spotting a group of low-slung dinghies than a single one adrift on the wide-open ocean. Sometimes the smugglers give the migrants a satellite phone to call for help once they're near or in the shipping lane. But if rescue doesn't arrive in time, whole boatloads perish at once. The name *death boats* is not only fitting but familiar: during the 1840s, the disgusting, disease-ridden vessels that transported 1.3 million Irish during the potato famine earned the name *coffin ships.*

Africans, who pay the least, get shoved in the ship's hold, when there is one. After a few hours at sea, these overcrowded, under-ventilated holds become sickening, slippery spaces. Everyone is piled on top of each other, puking and shitting together. People suffocate. And if the boat tips, everyone in the hold drowns. Syrians are expected to have more money and typically paid around $2,500 per person. More money meant more room and safety on board. Syrians travelled on the top deck or on less crowded boats, when they were still travelling the Libya-Italy route. One smuggler told Patrick Kingsley that he charged a Syrian family $100,000 for an inflatable Zodiac all to themselves.

In October 2013, a shipwreck off the coast of Lampedusa exposed just how deadly such a smuggling operation can be. When a boat carrying hundreds of mostly Eritrean refugees stalled a half mile from Lampedusa, the Tunisian captain started a fire on board to attract attention. The burning boat sank into the water. Tourists on the nearby beach thought the screams they could hear were seabirds. Three hundred and sixty-eight people drowned. Lampedusa once evoked a sultry Italian getaway; now it's a byword for refugee tragedy.

Public outcry prompted the Italian government to launch a search-and-rescue operation called Mare Nostrum (our sea), the Roman name for the Mediterranean. The support arrived at a critical time. In 2014, most migrants who crossed the Mediterranean used the route between Libya and Italy. Mare Nostrum sent five Italian navy ships, along with drones and helicopters, to patrol 43,000 square kilometres (27,000 square miles) of sea. The operation was a success: Mare Nostrum lifted 150,000 people, a third of them Syrian refugees, out of rafts and onto rescue boats. Some, including Britain's Foreign Office, expressed concern that Mare Nostrum was *too* successful. If smugglers could count on the Italians to deliver their human cargo, critics argued, this would encourage more people to leave and create an unintended pull toward Europe. The Italian government shut Mare Nostrum down later that year, citing high costs (around nine million euros a month) and a lack of support from the northern EU countries where the migrants actually intended to settle. In its place, the border agency Frontex started Operation Triton. With seven coastguard vessels covering the thirty nautical miles of water bordering Italy's shoreline, Operation Triton was a third of the size of Mare Nostrum. Its mandate was to secure borders, not save lives.

In the warmer migrating months of 2015, it became clear that Operation Triton was a failure. It could neither deter people from crossing nor save them when they did. During a single week in April, 1,300 people drowned. Either the smugglers in Libya hadn't gotten the message that Mare Nostrum would no longer rescue people at sea, or they didn't care. They had an endless supply of desperate customers willing to take a chance on a life raft. Nonprofits such as Médecins Sans Frontières and the Migrant Offshore Aid Station tried to fill the vacuum Mare Nostrum left, patrolling for boats outside Libya's territorial waters and covering as much seascape as possible.

At the same time, another crossing to Europe was gaining steam. In 2015, 873,179 people—nineteen times the number who crossed in

2014—boarded boats in Turkey, bound for Greece. Hassan Basheer was one of them.

❧

Like Mohammed, Hassan began his migration at a distance from the coast. In 2012, he wrote his last exam at Damascus University and he didn't wait to get his grades or his diploma. A few days later, he and two friends got in a taxi and drove the one hundred kilometers to Beirut, Lebanon. The Syrian army was on the prowl for young men, stopping them in the street, hassling them for studying rather than fighting in the civil war. On the day of his departure, bombs thundered in the distance—the sound of the regime laying into the nearby suburb of Al-Qabun. At each army checkpoint, Hassan was terrified that the officers might send him and his friends back to fight. Miraculously, they were let through. "I cannot imagine what happened after it got worse," he shuddered.

Before leaving Syria, his father tried to prepare him for refugee life. Hassan was born into a middle-class family—his father was an engineer, his mother a teacher—and he was raised in cosmopolitan Damascus. He dreamed of writing English-language poetry inspired by Arabic tradition, but he expected to become a teacher to pay the bills. He had never flown on a plane or crossed the sea by boat. He knew nothing about sleeping rough or sneaking across borders. Swept up in what the United Nations' International Organization of Migration has called the biggest human migration since World War II, he and twelve million other Syrian refugees were about to find out. Syrians weren't the only ones on the move. The events of the 2011 Arab Spring rippled through the region: mass protests, deposed leaders, economic downturns and widespread human rights abuses. Afghans, Iraqis, Somalians, Nigerians, Eritreans, and migrant Bangladeshi workers who had lost construction work during an economic crisis in the Persian

Gulf were all fleeing for a better life—a massive human exodus in an age of upheaval.

Hassan's father could never have fully prepared him for the turn his life was about to take: the danger, the boredom, the humiliations big and small, and the countless frustrating dead ends that marked his new life as an outsider in every country he landed in. His trip took him to the sandy no-man's-land between Egypt and Libya, where he slept on a border street, wrapped in his mother's bedsheets. In Beirut, he accidentally blew his devalued Syrian pounds on the most expensive hotel in the city. In Giza, he found an affordable apartment but had to flee Egypt after the new president, Abdel Fatah el-Sisi, labelled Syrians as terrorists. Finally, in 2015, he ended up in Istanbul. He was relieved to find a cheap room in a brothel. But when a john flung open his door and demanded to know where his whore was, he decided it was time to find a new place.

After Hassan landed in Istanbul, he started to learn Turkish. He found an under-the-table teaching job, where the school owner asked him to pose as a native English speaker so the school could charge higher rates. With his Western looks and fluent English, he mostly got away with posing as Dave from Seattle, but no one from North America would have been fooled. Not long after, the same unravelling that had happened to him in Egypt started again: he lost his job, the bills mounted, and his spirits spiralled downward. His Turkish students who knew him as Dave casually bashed Syrians in front of him. What would they do if they found out he was actually Syrian? He wondered what would happen to him if anti-refugee hostility flared up the way it had in Egypt after the military coup. Hassan continued looking for a new job, and it seemed like things were turning around when he found a better teaching job at a legitimate school. But again, the school owner asked him to pose as a Westerner and work under the table. In a country with close to three million refugees, cheap labor glutted the market. Very few Syrians received work permits or identity cards. This gave way to rampant exploitation and refugee child labor, particularly in Turkey's garment industry.

If he had left Syria any later, his situation would have been far more dire. Across the Middle East, millions of people played musical chairs as they looked for a safe place to settle. Temporary camps of displaced Syrians sprawled in neighboring countries. Years earlier, in the summer of 2012, the EU border agency Frontex had already recorded a 912 percent uptick in sea crossings from Turkey to Greece. The report noted that the boats kept coming despite the worsening winter conditions. The numbers shouldn't have been surprising. For years, people had crossed over land from Turkey and through Greece and Bulgaria. But these countries had recently built walls along their land borders with Turkey and introduced aggressive patrol guards. Bulgarian guards, in particular, had a brutal reputation for beating people who tried to pass through.

In the summer of 2015, more and more boats headed for Greece. The odds of surviving the journey were far better than they were on the much longer Libya-to-Italy route: only one in eighty-one died across all Mediterranean sea routes to Europe, compared with one in twenty-three between Libya and Italy. Many cellphones had reception out on the water between Greece and Turkey. These boats left at night. Unlike the reliance on passing ships along the Libya-to-Italy route, migrant boats in the eastern Mediterranean had to evade the Turkish and Greek coast guards—unless the boat started to sink. Hassan's friends suggested joining the exodus. Feeling disillusioned and worried about his future in Turkey, he made a snap decision to leave. Going to Europe was not the first step. It was the last solution.

❧

As Mohammed's boat set out for Italy, the captain sang a farewell song to Libya. Mohammed was not sad to see the last of that country drift away. But as awful as Libya had been to him, it was still difficult to say good-bye to everything he knew. The coastline receded farther and farther away. Suddenly Mohammed was facing the open sea, without a life jacket and unable to swim. As demand grew and smugglers looking to make a quick buck oversaturated the market, the price for a single passage went down and smugglers began overloading the boats. The dinghies were built to carry between twenty to thirty people at most, but Mohammed's dangerously overloaded boat carried over a hundred souls. He had only a few snacks to last what could turn into an unpredictably long journey. He was absolutely vulnerable to the water, and it was terrifying. A few minutes passed before Mohammed saw the second boat following his own. He searched for his friend on board, but he couldn't find him among the hundred or so faces.

"As we moved deep inside the sea, that's a place I'll never forget in my life," he said. "I've faced many things in Ghana, in Libya, but that's the main thing I'll never forget." He heard some sort of commotion on the second boat and looked back. Whatever happened, happened fast. People were scrambling, and then everyone was in the water, thrashing and sinking into the sea. His own overloaded boat could do nothing. Helplessly, Mohammed listened to the sound of a hundred people drowning at once. He felt certain that his friend was among them. Eventually, the screaming stopped.

Later on, his own boat ran out of fuel. As they drifted on the Mediterranean, Mohammed's fate rested entirely on the whim of the sea. How long would the fair weather last? Would a passing ship spot this dinghy all alone on the ocean? He saw dolphins leap past and some other fish he didn't know the names of. Everyone tried to sit as still as possible, not wanting to strain or capsize the fragile boat. Time passed. Then he saw the floor of the dinghy starting to deflate beneath him. Salt water seeped up around the flimsy deck. Screams and cries rose up around him, and Mohammed began to cry as well. They'd seen what had happened to the other boat; they knew what would happen to them. Somehow in the melee, one woman imprinted on his memory. She was young and pregnant, and she had a young child with her. She was the only woman who didn't cry.

As the small dinghy headed out to sea toward Greece, Hassan checked the time on his phone. It was 2:00 A.M. He waited and checked again: 2:01 A.M. He waited some more and checked again: 2:02 A.M. Every moment felt interminable on the water. Every muscle in his body was clenched and waiting for something horrible to happen. His family in Damascus knew that he planned to cross the Mediterranean, but he'd decided not to tell them exactly when. Only his brother in Istanbul knew he'd left. He had nothing to do but check the time and check the time and check the time once more.

His destination was Chios, an island two hundred kilometers north of Kos. In just a few months, Alan Kurdi would drown while trying to reach Kos. There were no seats on the boat. All forty-five people on board simply piled on top of each other. Hassan ended up squished in the center of the dinghy, where the women and children usually huddle. The men typically stand on the periphery of the boat, as if on guard against the sea, except the strategy often backfires. The center of the boat, loaded down with more passengers, fills with water first. Survivors tell stories of women who hold their babies above their heads as the boat is sucked down at its center. The life jacket the smugglers gave Hassan looked suspiciously like foam and not remotely floatable. He was sure that if the boat sank, he would go down with it.

With so many people piled on top of him, Hassan quickly lost the feeling in his legs. The boat rose up and down in erratic, nauseating jumps. A child beside him had a puckered look on his face, like he was about to puke. Hassan tried to coax him out of it. He checked the time on his phone again. Every moment dragged. He had entered the space in between life and death that the philosopher Anacharsis once described sailors inhabiting at sea, their lives hanging continually in suspense in front of them until they reached land. Hassan was not dead, but at that moment he was not really alive, either. Until he set foot on land, his life belonged to the sea.

He checked the time on his phone again. "What's that light?" people yelled. They were scared that even the tiniest beam might attract the Turkish Coast Guard, which would force them to return to shore. Hassan snapped his flip-phone shut. It instantly lit up again, pinwheeling bright lights into the darkness. It was his brother, back in Istanbul, calling to see if he had made it to shore already. "Turn that phone off!" people screamed at him.

From his position in the dinghy, Hassan couldn't see the ocean. It was disorienting to be surrounded by water: to hear its shattering sweep, to feel its coldness, to be aware of its presence pulling the boat up and down over the waves but to see only a crowd of heads and the black of the night sky above. He wondered whether they were moving in the right direction.

The drivers of these boats were not chosen for their seafaring abilities. They were refugees who volunteered in exchange for free or discounted passage. Sometimes they got an hour or two of practise with the motor before leaving. They travelled blind, with the simplest of directions from smugglers: go straight, turn left at the light, something like that. This could go wrong in so many different ways. Boats chased after the bright lights of a cargo ship or became disoriented in the darkness. This didn't build a lot of confidence in the "captain" among the passengers. Boats sometimes devolved into shouting matches, backseat directions hurling through the air. Luckily Hassan's driver came from a coastal town in Syria, and he knew how to pilot a dinghy. But still, the conditions would have challenged an experienced boater and voices floated over the dinghy, arguing over what the captain should do. The tiny outboard motor struggled to push the dinghy up a huge wall of water. The boat slid down, out of control, into a deep trough, and then it was up another wave and down into another trough. Meanwhile, the wind and the waves buffeted the dinghy farther and farther away from the island of Chios.

Hassan's feet felt cold. He looked down. The water was at his waist. The boat was sinking, and quickly. Finally, after days and weeks of waiting onshore and wondering whether this would happen, it was reality. He was face-to-face with his greatest fear: drowning in dark chaos. It would make no difference to the ocean whether he lived or died, but Hassan desperately needed the ocean, this indiscriminate place, to spare his life that night.

His brother would have to tell the rest of the family what had happened to Hassan. Standing at this border between life and death, the risk he had taken became crystal clear. As the water rose around him, his vision of a new life in Europe receded. Hassan had heard stories of people screaming as the boat went down. But no one called out or wept. There was only silence, interrupted by the occasional whisper and prayer.

Just as Mohammed's boat began to deflate on the water, a ship appeared on the horizon. As the ship circled close, his rescuers spoke to the group in English. The ship would send over a skiff, the crew explained, and rope the two vessels together so that everyone could board the ship. But first the rescuers collected a three-month-old baby from a mother's arms on board. Everyone would board one at a time, women and children would go first, followed by the many men. As the boat started to sink into the sea, the mood on board became agitated and anxious. Safety was so close, yet if the boat tipped at that moment, dozens could still die.

The worst shipwreck in modern history had occurred on the Mediterranean only a year earlier when a trawler transporting hundreds of migrants flipped. Dozens of people spilled off the deck and into the sea. They were the lucky ones. Trapped below deck were hundreds more who had no hope of swimming through the cabins and up to the surface. Eight hundred and fifty people drowned. Ironically, this happened during a rescue operation, and photos of similar rescues gone tragically wrong appear online, freeze-framing the second migrant boats capsize. Rescue at sea is a dangerous operation, even more so when you add in the tendency of desperate people to rush for the exits as soon as safety is in sight, the dangerously or improperly loaded boats, and the unskilled and inexperienced captains. Everything conspires to throw off an already precarious balance.

As the skiff pulled up next to the sinking dinghy, the inequality between the vessels became clear. The skiff was empty except for a few crew wearing helmets, life jackets, and sunglasses. The other was packed to capacity and beyond, its floor covered in garbage and the people exposed to the elements. The inflatable dinghy was roped up next to the rescue ship and a staircase descended for the rescued to climb aboard. As Mohammed climbed up, he registered the English accent of his rescuers. Someone from the crew hustled him off to another part of the ship to replace his ripped pants. The ship sailed north for another day where Mohammed was transferred to a German ship that took him the rest of the way to Italy. On land, he was brought to a holding compound, where he slept outside and received nothing to eat. The authorities led small groups away and asked them to write their personal information on a slip of white paper. He didn't have a passport on purpose. Passports were "dangerous" he said, because they prove his home country and make his deportation even easier. Before he was interviewed by the Italian authorities, Mohammed left the holding compound with a group of Africans headed for Germany. On the train north, he learned how to navigate through Europe from the other migrants on board. He knew to tell the border guards in Switzerland that he planned to stay in the country, or they would send him back to Italy. The guards gave him a ticket for a place to sleep at a campus nearby. Instead he purchased a train ticket to Germany and finally, after miles by sea and sand, arrived at the borders of a country he hoped would keep him safe.

~

After ninety long minutes on the choppy water, Hassan's sinking boat approached a dark island. Huge rocks reared up out of the black night. Waves crashed off them. The migrants on his boat were terrified of landing there, but the boat was half-full of water and the sky was threatening a storm. It would have to be an emergency landing or sinking at sea. An emergency landing can be a fatal way to end a crossing. If the boat flips as it beaches, people are tossed onto sharp rocks, where the waves knock them down or drag them back out to sea. Hassan was lucky. His captain timed a perfect landing between the waves, gliding the boat up and over the rocks. Hassan stepped out. Never in his life had hard rock beneath his feet felt so good.

After everyone climbed out onto the rocks, one man knifed the boat to shreds. Rumors circulated about the Greek and Turkish coast guards pushing migrants into their boats and back out to sea right after they landed. This treatment became so common it acquired a name: pushback. In 2009, the Italian government intercepted a boat of two hundred migrants near Lampedusa and returned them to Libya. During the spring of 2015, Thailand, Malaysia, and Indonesia pushed boatloads of Rohingya asylum seekers back out to sea and into the Bay of Bengal. It becomes difficult to do that, however, if the boat won't float. One Afghan teenager I interviewed said someone on his boat punctured the vessel before it reached land. Everyone on board was forced to swim the last forty meters to the beach.

Hassan looked up at the deserted island and realized this could not possibly be the resort island of Chios. He saw not a single structure on the island except for an abandoned church on a hilltop. They had arrived on a small Greek military outpost called Pasas, which lies northeast of Chios. Only later did he find out where exactly they had landed. At the time, he hoped that he was at least standing on European soil.

Of the three boats that left Izmir that night, only two arrived at Pasas. Someone told Hassan that the third boat turned back, but he didn't know whether or not that was true. Moments after they stepped ashore, cold rain came down in sheets, and high waves crashed onto the island. They had made it to land just before a storm. It was a cold, miserable night to stand unprotected on a lonely rock of an island. The abandoned church offered the only shelter. A few people started to gather at its boarded-up doors and discuss breaking in. Hassan worried about how this might look to the authorities—a group of refugees breaking into a church as soon as they made landfall. The soldiers who watched over the island arrived before the group came to a decision.

As the young Greek soldiers walked up, it dawned on Hassan how desperate his group looked. This soaked and chilled group of Syrians, standing in the rain on a deserted island, the children shivering, their lips turning blue, everyone arguing about whether to break into a church. He saw panic in the soldiers' eyes. Then he realized they were scared *for* the refugees, not *of* them. The soldiers' humanity humbled Hassan. He also felt incredibly ashamed that his country, as well as the world, had forced him out on the water that night.

❧

In a German village near the Swiss border, a group of teenagers and social workers sang a Gaelic blessing. The song was translated into German, but the dominant language in the room was Farsi, followed by Somalian, Pashtun, Arabic, and African languages like Hausa and Fula. The teenagers were dressed in an all-black street style, and they mumbled through the chorus: "*Und bis wir uns widersehen, halte Gott dich fest in seiner Hand.*" A tall, blond social worker sang loudly to make up for the muddied lyrics.

Mohammed had arrived at this group home a month earlier and had sung this farewell song dozens of times without understanding a single word. In the coming month, he would hear this song for the last time when he left the group home for a communal apartment in a nearby town. He would also go through numerous interviews with social workers as they tried to determine whether he could stay in Germany. During his first interview, he wanted to be absolutely truthful, so he admitted to smoking marijuana sometimes. When Mohammed saw news reports on TV of people floating on rafts in the Mediterranean, his memories swept over him. He would excuse himself, find a quiet place to smoke, and try to calm down. During his stay at the group home, a social worker was surprised to learn that he didn't know how to swim and offered to teach him at nearby Lake Constance. He refused. While the other migrants learned to swim, he sat on the shores and smoked. The thought of his feet kicking, unable to find ground beneath him, was unbearable. One day, as I was sitting with him on a park bench near his group home, he went back to the worst moments of his life, something he rarely let himself do. Out on the water, far from land, watching "a hundred plus people losing their lives . . . ," he said, trailing off. The silence between us widened into a minute, and then five. He cracked his knuckles, stared at the ground. Tears built in his eyes. Eventually, he regained his composure. "It's not easy," he finally concluded. I was the second person to whom he had told the story of crossing the Mediterranean. The social workers at the group home steered away from asking unless someone wanted to share.

An hour away from Mohammed's group home, Hassan Basheer had a flat of his own in a village on the shores of Lake Constance. He and Mohammed have never met. They'd find it entirely unsurprising that their experiences mirrored each other's but from opposite sides of the Mediterranean. Almost everyone they knew in Europe was a refugee like them with a horror story to tell.

One night during Ramadan, I sat with Hassan at a restaurant near his home. I was starving, but he was fasting. I took guilty bites of bread whenever he looked away. He told me about his time in Izmir and imitated the Turkish shopkeeper who sold him balloons to protect his cellphone from the ocean. "This, sir, is the best balloon," Hassan said, puffing out his chest like an overly confident car salesman. Hassan had an eye for the absurd. Laughter felt like a way to claw back power after the dehumanizing crossing. Along the journey, refugees are treated like animals: crowded into filthy *mazraas*; wrapped in bedsheets on a dusty street; crammed onto a dangerous boat without food, water, toilets, or sometimes even ventilation. On his journey, the way Europeans treated him often humiliated Hassan, like they were processing a contagious animal. The subhuman handling continued on land: politicians and media talk about refugees as a herd, a scary other, or a natural disaster, a *flood* of people *swamping* the country's borders. It's as though a refugee who claims asylum is an attack on a state's nationhood, even though Hassan and Mohammed were the ones who had truly lost their homes.

The refugee's right to claim asylum is a rare chance to rewrite the story one was born to live. The 1951 Refugee Convention promised a watershed moment in defining that right, one that today is in danger around the world, and even more so when refugees arrive "illegally" by boat. "Refugees and others who migrate by boat have since been considered the most threatening," writes scholar Lynda Mannik. "The 'official' reason behind this [threatening] designation is that their movements are uncontrolled and often uncontrollable and that their choice to migrate illegally is a criminal act of sorts, whether or not they are involved with smuggling rings."

In Germany, politicians are under pressure to cut refugee numbers, and the definition of a refugee is not as inclusive as it should be. A Syrian refugee like Hassan passes muster, but a gay African teenager's path is less clear. Technically Mohammed should qualify under the 1951 Refugee Convention because he crossed borders and is unable to return to Ghana, where he has a reasonable fear of persecution. Mohammed doesn't know why his claim was rejected, but the executive director of the European Region of the International Lesbian, Gay, Bisexual, Trans, and Intersex Association (ILGA) recently noted that the success rate of LGBTI asylum applications in Europe has dropped. Germany classifies Ghana as safe; Mohammed is dubbed an economic migrant and deported.

Economic migrant is another blunt label that disguises the complicated past of the slave trade and decolonization in Africa. Some of the richest nations in the world, including the United States and former European colonial powers, are becoming more hardened against economic migrants. Ironically, the United States and Western European powers profited from and interfered with Africa's development throughout the 20th century: exploiting resources, deposing democratically elected leaders, installing dictators, and hampering the continent's ability to self-govern. No wonder this continent faces a unique set of challenges as it develops healthcare, education, economies, and infrastructure that match those of the developed world.

None of this history is acknowledged when an African washes up on the shores of Europe. Many of the millions of Europeans who migrated by boat in the 19th and 20th centuries were celebrated for pursuing a better life outside their home countries. The Africans now looking for opportunity on the continent that has taken so much from them are not admired. Instead, depending on whether their countries of origin are considered safe, they are classed as an economic migrant, a grasper and a threat.

Hassan says that the majority of men who left Syria were fleeing military service, like he was. It's not that I was afraid to die back then," he said. "I was afraid to kill." Starting over in a new country is a lonely life. But they were luckier than those left behind: women, children, the sick and elderly, fending for themselves in dangerous or diminished countries. Whole families pin their future on a son or a brother or a husband, hoping he'll send money back or bring them over once he's settled in Europe.

Over the years, Europe has tried to reassert control over its coastline and address the push and pull factors that bring asylum seekers to its shores. Fences are a growth industry in many coastal European countries. Greece and Spain have spent over €70 million on boats, drones, and off-road vehicles to police the coast. Since 2011, Italy has given seventeen million euros to the Libyan Coast Guard for boats, training, and night-vision goggles—whatever it takes to stop people from crossing the sea. Germany has brokered aid deals between the EU and Turkey to take better care of the refugees inside its borders, like educating Syrian youth and allowing Syrian refugees to work legally in Turkey with a permit. That policy might have stopped risky journeys like Hassan Basheer's back in 2015, but so far the integration of Syrians into Turkish society has had mixed results. The resentment that Hassan once feared rising against Syrians when he lived in Istanbul seems to have arrived amidst a declining Turkish economy. Recent reports speak of attacks against refugees and forced deportations to Syria.

The ocean remains one of the best tools for keeping unwanted people out of a country. Island and coastal nations, such as Canada, Japan, and New Zealand, can cherry-pick the nationalities and skills they want to let in. The United Kingdom can stop people crossing at the English Channel, where underage refugees once waited in the overcrowded, now demolished, Calais Jungle. Hassan Basheer and Mohammed Botwe knew the risks of crossing the sea. They boarded boats for Europe anyway. Only two options lay before them: survive the ocean or die trying.

Mediterranean crossings have dropped, but they haven't stopped. People continue to leave for Europe, and building in deterrents only makes the crossings more dangerous. Migrants continue to cross in the highest numbers between Libya and Italy. Since 2015, the European Union has diverted funds to the Libyan Coast Guard, a notoriously corrupt, poorly trained, and fractured organization, but notably one that will return migrant boats to Libyan soil rather than bring them to Europe. According to Reuters, the EU and Italy plan to spend forty-four million euros on training and outfitting the Libyan Coast Guard until 2020 to help rescue boats beyond their territorial waters. Human rights organizations and activists have denounced the EU for handing over rescue responsibility to Libya, which takes migrants to inhumane detention camps. Libya has established five twenty-four-hour phone numbers for boats in distress. A recent investigation by Buzzfeed found that not one of the five numbers was answered with any regularity. The rescue ship *Sea Watch 3* tried to call fifteen times; ten of those calls failed, and five times the Libyan Coast Guard answered and then hung up.

FOUR

FLOATING FREE

The greatest offence against property was to have none.

—E. P. Thompson

Daniel Inkersell is a handsome, cheeky man who likes to have a time, as he says in his bouncy East Coast accent. A gregarious twenty-six-year-old like him is typically in the city—dating, drinking, building a life. Instead Daniel tied his future to a crumbling pillar in the Dogpatch, a motley collection of off-grid boats moored in the harbor of a seaside town on the coast of Vancouver Island. When a bylaw officer pulled up to Daniel's boathouse one morning, Daniel was inside, drinking a cup of coffee and contemplating that morning's boat repairs. The officer handed him an eviction notice, and Daniel, trying to keep it professional, asked the bylaw officer for more details about the infraction. His boat, he was told, had violated the zoning bylaw that restricted anchoring in the water lot to seven days. Daniel had a month to move. If his boat remained after the eviction date, the town could seize his vessel without notice, leaving him no other option but to return to land.

Ladysmith is a small, bucolic seaside town with a low crime rate, an aging population, and a picturesque downtown that's cut off from the water by the highway. The Dogpatch is literally on the wrong side of the tracks. To reach it you have to cross that highway and the overgrown railroad tracks before descending a forested path to the waterfront. According to many townspeople, the boats anchored in Ladysmith's harbor are "eyesores" that leak oil, and the people who live on board are "water squatters" who clog the harbor, discourage tourism dollars, and bring a slew of unseemly problems to otherwise idyllic Ladysmith. Rumors swirl that at night thieves come to shore to pillage, and drug dealers to ply their trade.

On land, there's usually a process for establishing the owner of a troublesome property and holding that person to account. On the coast, it's murkier because no single person owns the water. The ocean is a natural national border to the county and a shared resource, so it's under the jurisdiction of Canada's federal government, as it is in many coastal countries. Anchoring a boat is part of navigating, which also falls under federal jurisdiction in Canadian law. Boat owners in the Patch contend that they're not trespassing on public water but anchoring legally.

This slippery definition of navigation in maritime law allows poor and vulnerable people to live free on the water. Boaters living on the water also argue that if a vessel serves as their primary residence, taking away that home or pushing them out of a safe harbor during the stormy winter months contravenes Canada's Charter of Rights and Freedoms. The ocean is one of the last refuges in the developed world where the poor, the homeless, and those who simply prefer an independent lifestyle can live under their own shelter. The town's homeowners want cleaner oceans and waterfronts that operate like a public park.

The Dogpatch doesn't fit with Ladysmith's plans for itself as an eco- and tourist-friendly retirement community. The Patch residents—a mix of adventurers, rebels, and societal misfits—live off the grid, and unlike the roughly twenty boats moored at the nearby marina, they don't pay for moorage. The Dogpatchers know what people say, but to them it's just mudslinging. They've endured far worse.

As the bylaw officer motored away to tape eviction notices on the other boats anchored nearby, Daniel considered his options. The nearby boatyard had kicked him out for repairing his yacht, *Sojourn*, on land, so he'd brought *Sojourn* to the Patch out of necessity, not choice. He'd taken the boathouse that had come with *Sojourn*, floated it out, and roped it to four concrete pillars (remnants from the harbor's industrial past) to house his 48-foot-long Chris Craft yacht. That had not been easy. An affinity with wooden boats runs deep in the Inkersell family. His dad once owned a sleek, all-wood Chris Craft similar to *Sojourn*. His grandfather, a wooden-boat surveyor, dismissed any vessel made of fiberglass as Tupperware.

Daniel's plan was to refurbish the $8,000 yacht himself with money he'd earned planting trees, picking grapes, fighting fires, and ruining his back, and then upsell her to a Chris Craft fan. He still had many more repairs to make before he could get a return on his investment. Before he'd found this permanent spot to anchor in the Patch, he'd spent his nights sleeping in his truck and varnishing the boat wherever he could. Sometimes he'd find himself in the Patch, harnessing *Sojourn* to a makeshift buoy and rolling in the rain. Finally, two months earlier, he'd found this spot, and he'd been making progress on *Sojourn*'s engine. Then, with a sputter, the bylaw officer pulled up, brandishing an eviction notice. Reaching a compromise between the town of Ladysmith and the Dogpatch seemed easier to Daniel than cutting loose for open water. If he motored off, would he find anywhere else to accept him?

❧

Water-dwelling communities come in all shapes and sizes, molded by the coastline, the climate, and the culture surrounding them. Water nomads called the Bajau once travelled and lived on the waters of Southeast Asia in teak houseboats. In Amsterdam, *woonbootbewoner* live on canal boats or float homes that range from run-down and rat-infested to sleek and modern. In Sausalito, California, cute and colorfully painted float homes line Richardson Bay where Otis Redding supposedly penned "(Sitting on) The Dock of the Bay" while he stayed in one float home. Just below the Arctic Circle, a community of liveaboards hack out a living from the frozen shores of Yellowknife's Great Slave Lake. For a time, you could watch their goings-on on the short-lived reality show *Ice Lake Rebels*.

Across these diverse communities, a few traits hold true, at least in the West (it's harder to make generalizations about the Bajau because the communities are more scattered and, according to researchers, stretch back at least a thousand years). These communities grew out of the freewheeling days of the '60s and '70s. They're more offbeat than your typical neighborhood on land. Living on the water draws in a fringe group: eccentric, creative, handy types, such as inventors, artists, engineers, and fishers. They take pride in catching their food, rebuilding their engines, and living outside the mainstream.

Looking back throughout history, the port was a crowded, busy place of industry where people of different backgrounds, cultures and incomes mingled and traded. Many poor people lived near the waterfront because the living was cheap and many of them worked there, too. In 19th-century Seattle, loggers slid their timber from the mountain along a road to the mill on the waterfront, giving birth to the term *skid row*. Over time, skid row became shorthand for the poor, crime-ridden areas that often clustered at water's edge.

In San Francisco is another, more famous Dogpatch that was also once a gritty waterfront district populated by people who migrated there during the Depression, looking for work in the shipyards. "This was during the time that Li'l Abner was a popular comic strip," Christopher VerPlanck, an architectural historian, told the *New York Times*. The strip took place in a fictional hillbilly town called Dogpatch. "San Franciscans saw the Lower Potrero area as a working-class, white, Southern enclave and called it Dogpatch."

In post–World War II Amsterdam, working-class families who couldn't afford housing on land bought barges retired from the Dutch navy and took to living on the city's canals. In her Booker Prize–winning novel *Offshore,* Penelope Fitzgerald captured a similar world in the downtrodden life of water-dwellers along the Thames River in the early 1960s. "The barge-dwellers, creatures of neither firm land nor water, would have liked to be more respectable than they were," Fitzgerald wrote. "But a certain failure, distressing to themselves, to be like other people, caused them to sink back, with so much else that drifted or was washed up, into the mud moorings of the great tideway."

But as cities deindustrialized and developers converted former shipyards, docks, and factories into tourist destinations, living near the urban coastline became desirable. "For a long time, people didn't want to live close to the water, but you certainly see a lot of valuable real estate today on the shoreline," said Andy Yan, an urban planner who studies high real estate prices in Vancouver. "We've come from a utilitarian relationship to an aesthetic one." Today, real estate agents can tally in dollars and cents just how much that ocean view costs. A 1997 survey in Point Roberts, Washington—a tiny US peninsula in the Strait of Georgia accessible by land only via Canada—found that a house with an oceanfront view could cost up to 147 percent more than a house without one. A 2010 study in South Carolina pegged the waterfront premium as high as 287 percent, although that premium may decline in the future with the prospect of climate-change induced sea level rise. One Finnish study found that even the quality of the water is a commodity, and people will pay up to 20 percent more to live near cleaner coastlines. On the Thames River, homes now sell for over a million pounds. In San Francisco, Dogpatch condos cost hundreds of thousands of dollars, although some buildings are Airbnb-friendly for those who can only afford a night there. At the San Francisco airport, you can buy a ten-dollar sandwich at the Dogpatch Bakehouse & Caffe. In Amsterdam, the canal boats and float homes are no longer the last resort of cash-strapped families; they're now cute bed-and-breakfasts for tourists.

Environmentally, the waterfront's gentrification has been an unmitigated force for good. Where effluvium was once flushed without second thought into the ocean and pools of creosote floated, we now talk of raising the water quality and collecting plastic. Amsterdam's canal residents were required to pay for an expensive upgrade that connected their boats to the city's sewer system. Now, in the water between the homes of the Ijberg neighborhood, children horse around on swimming rafts and pool noodles. The Billion Oyster Project aims to cleanse the once lifeless New York Harbor with millions upon millions of water-filtering bivalves. In Copenhagen's formerly polluted port, children learn to swim, and pool parties flourish on warm summer days at the harbor baths.

More people bring more competition for space. Another trait that runs through many liveaboard marinas and water-based communities today is a sense that their days are numbered. From Fort Lauderdale to Hong Kong to Vancouver, marinas are being squeezed by oceanside development and the rising cost of living. Waiting lists stretch a decade long, slip prices are on the rise, and marinas engage in renovictions: kicking out long-term tenants to upgrade the docks and then up the rental price. As the cost of living legally at a marina creeps upward, it becomes comparable to life on land, not to mention the surprise costs that come with maintaining a boat. As the available space continues to shrink and prices continue to rise, who gets to enjoy the water will be even more fiercely contested.

Municipalities tend to like offshore sailors (like Fiona McGlynn and Robin Urquhart), who pay to stay a few nights and move on. They're less enamored with people like Daniel Inkersell and the residents of the Dogpatch—folks who take up residence on the water and hold on as long as they can when they can't find a slip or afford its asking price.

❧

When Daniel first anchored in the Dogpatch, he heard a funny clink-clink noise pinging off his anchor. At low tide, he shone a searchlight into the murky water and saw that his anchoring line had snagged on a sunken sailboat resting on the coal-blackened bottom of the Dogpatch. Another resident of the Patch estimated that beneath the surface lay hundreds of derelicts, jettisoned by nearby marinas or neglectful owners, farting up fuel and that distinctive gasoline sheen. "The Dogpatch is like this big red dot on everyone's map," Vince Huard said. A few years ago, bowing to public pressure, the Canadian government counted all the derelict boats in its western waterways. The Dogpatch had the most with forty-five boats, some abandoned but many of which were lived on. Before he took a job as a fisher, Vince worked for a local business that disposed of derelict boats people dumped here. That was how he found the one he lived on in the Patch.

Time is slippery here. Vince guessed that he left the nearby marina to live in the Dogpatch four years earlier. "We all kinda forget the day. And I don't have a mind for dates," he said. Certainly, he arrived before the Dogpatch grew into a place the regulars called home. The community evolved at an infinitesimal pace. Sometime in the late '90s, the first resident of the Dogpatch, a veteran named Paul

Coop, arrived, and more soon followed, realizing they could live on the same water as the nearby marina but for free. At first, few stayed permanently. As one boat left, the next took its place. After a while, however, people started to stay. They had to haul their own water and fuel, but that suited the type who dropped anchor in the Patch. They were pensioners, welfare and disability recipients, sailors, and shipyard workers, but all of them were people with more time than money and a commitment to life outside the mainstream.

Ironically, Ladysmith's beginnings as a coal-mining town drew in this community and gave birth to what many considered the town's most pressing problem. At the beginning of the 20th century, workers used to wash coal right next to the present-day Patch. All the toxic black dust they washed off coal—the slack—grew into a thirteen-acre triangular spit. The locals called it Slack Point, and it formed a perfect breakwater against the prevailing winds that swept down the channel in winter. When the coal mine shut down in the 1930s, this protected stretch of coastline housed a log-sorting operation. In the 1980s, the log-sort closed, and the town talked of finally cleaning up the waterfront.

Too many obstacles stood in the way. Environmental remediation costs for the coal-tainted banks were estimated at $27 million. The municipality and the water-lot owner disagreed about who had control over the shore, and the banks of Ladysmith's harbor sat undeveloped. Townspeople walked their dogs along the coal-black banks of Slack Point. The sea life crept back: writhing nudibranchs, purple starfish, gray Dungeness crabs. In the mid-1990s, abandoned boats began dotting the surface of the water, and people started living on those boats or bringing their own. "I saw that it was safe water, that it was protected," Bryan Livingstone said when I asked him what brought him to the Patch decades earlier. Before Bryan retired, he made his living chasing down lost logs and bringing them to the still operating log-sort. To him, it made sense for him to live in the Patch, and he is one of its longest residents.

"Most people think that the water is something that's supposed to be beautiful, but it's not," Bryan explained. "Navigable waters are the frontier of a country that interfaces with a thing called the ocean." Before the 1990s, there weren't too many other people in the harbor to argue with him. Now Ladysmith wants to integrate its sliver of ocean more directly into the town, just as other cities around the world have done, by turning it into a picturesque destination that draws tourists and earns money. It might want to commodify Ladysmith's coal-mining, log-sorting heritage, but it doesn't want people like Bryan, whose way of life references the hardscrabble reality of that history. "I think the whole situation down there needs to be cleaned up," one local told the *Ladysmith Chemainus Chronicle.*

According to one persistent rumor in town, the Dogpatch residents row to shore at night to rob the seaside homes. The local paper ran regular front-page stories about the nefarious misdeeds in the Patch: all-night parties, drug dealing, vendettas settled by arson, shouting matches, theft, and, perhaps worst of all, the dumping of bodily waste. At a town hall meeting after the eviction notices went out, an elderly resident complained about the Dogpatch. What happened to their sewage? Why were they allowed to sprawl into the traffic lane? "I don't have a dog in this fight," the man insisted. "I just don't want to see our beautiful harbor ruined."

Distrust between land- and water-based communities has a long-running history, stretching back to the time of Plato and Aristotle. Throughout history, agrarian societies that lived farther inland were often thought of as safer and honest, where people made their living off hard work and an enterprising spirit. The harbor was where deals were cut and fortunes made or lost, where different races and cultures mixed, and where incoming ships and loose morals spread disease and pestilence. Plato suggested that cities be situated at least fifteen kilometers inland to avoid the inevitable taint of the harbor; Aristotle believed that sailors need not be citizens with voting rights in early Greek democracy. But something deeper lay behind this age-old division that sparked the town's dislike: the Dogpatch was rootless, and the people who lived in it couldn't be held accountable the way they could on land. The Dogpatch liveaboards had antiquated cellphones (when they had them at all) and no mailing addresses. They didn't pay property tax. Their boats were isolated refuges adrift in a modern world where most of us live highly documented existences with long lines of credit history and updated social feeds. From the transient Romani people in Europe to the "pikey" slurs against Irish travellers, the mainstream often sees nomadic groups without local roots or connections as a threat. Indonesia and the Philippines have since recognized Bajau water dwellers as citizens, but this has come at a cost to the culture's transient history. They've mostly traded their teak houseboats for stilt homes built a few meters above the ocean and live in sedentary communities that have a financial relationship with nearby towns and villages.

It's hard to argue with a town's desire for a healthier, more visually appealing waterfront. But I still felt drawn to the boat-dwellers' perspective. It surprised me that two groups of people who lived parallel lives could see the ocean in such fundamentally different ways. When I heard the town residents complain about the Dogpatch's lack of sewage treatment or their sprawling boats or their tax avoidance, what I heard was: we can't trust them. That distrust cut both ways.

A decade earlier, the government of British Columbia made its first attempt to clear out the area and hauled away docks and equipment unannounced. The event still haunted a few Dogpatchers who came home to find their few possessions missing. By most accounts, the Patch cleaned up after that first sweep, and a relative peace took root.

Until the fires started.

During the summer of 2015, a forty-foot boat caught flame in the night. All night, Dogpatchers and Ladysmith's volunteer search and rescue team, aided by the Coast Guard, worked to keep the flames from spreading to nearby boats. Police deemed the fire "suspicious." Later that summer, two boats caught fire—one burned to the waterline, and the other was left to drift ashore. The RCMP wondered whether vigilante justice was spreading across the water.

❧

Daniel Inkersell began to spend his days motoring from boat to boat, collecting signatures from people in the Patch. He had written a letter to Ladysmith's mayor that pleaded for reconciliation and time before the eviction date. Daniel didn't own a cellphone, so he couldn't call his neighbors. Communication in the Patch was, well, patchy. The wireless reception was steady, but many of the dozen people who lived there didn't have cellphones or email accounts. News spread from ship to ship or ship to shore, and the people used the surface of the water as a natural megaphone.

Everyone in the Patch was at work on a half-finished project. Daniel labored over his engine. Vince Huard was fixing up a second boat he had salvaged and chained up next to the boat he lived on. Lew McArel repaired a mountain bike in the rain. These projects were a proxy for larger life goals: to earn money, to become self-sufficient, or to simply explore. Living on the water afforded people the time and luxury to learn a new skill. "Part of the reason I bought this boat is not only to refurbish it but to develop myself as well," Daniel explained. "Experience, the way I see it, is knowledge. When I'm fifty, I don't wanna be this guy who doesn't have much life experience and is in this shitty job." For most, the Dogpatch's appeal wasn't the free rent; it was the identity.

In the Patch, no one was impressed if you paid for something new. But if you found it and jerry-rigged a solution—now, that was a story. The Dumpsters near the shore acted as an informal trading post. A composting toilet materialized; a few hours later it was gone. One man told another he'd dropped off a TV at the Dumpster but forgotten to include the remote control. "Oh, I picked that TV up," the other man said. They exchanged the remote control.

When the Dogpatchers heard residents complain about their environmental impact on the water, they pointed out that the town's wastewater ran downhill to the ocean, an echo of the town's earliest days when miners washed coal at the shoreline. When it rained, oil drippings from Ladysmith's streets congregated in a slurry at the shoreline. The town's treated sewage also went into the ocean, and let out not far from the Dogpatch.

"We in the Patch are not the contaminators," Bryan Livingstone stressed. "We are living on the leftovers, the discards. We have, ecologically, a very small footprint." This commitment to reuse and reduction was one of the main points Patchers used to defend their lifestyle. Perhaps if the Dogpatchers contributed more to the local economy, supporting jobs and small businesses with their dollars, it might have created a feeling of exchange and rapport between the two sides. But instead of shopping at the touristy stores along Ladysmith's First Avenue, the Ladysmith Health Care Auxiliary Thrift Shop was the store of choice for the Patch. One man said he scoped out the firsthand stores on Ladysmith's main drag to see what he would be wearing in a few years, bought on the cheap from the thrift shop. Lew McArel had no problem with wearing women's jeans if the zipper was long enough. A few Patchers kept track of where the restaurants and grocery stores off-loaded their goods at the end of the day. At night, they dove into Dumpsters, hauling back bent vegetables, day-old buns, and wooden pallets to burn in their stoves. This might have been what started the rumor in town of Patchers pillaging at night. But these men called themselves "freegans"—a half-truth mixed with a bit of pride—and their gold was crusty cinnamon buns.

Some townspeople admired the Dogpatch's tenacity. "I personally consider most of the liveaboards to be the stewards of the Dogpatch," read one letter to the editor of a local magazine. "And I say: Let them stay." Ladysmith's mayor, Aaron Stone, also seemed sympathetic. He wanted to find a way to accommodate the Patchers. His main goal was to remove an abandoned 108-foot trawler called *Viki Lyne II* from the Patch.

Ladysmith's local newspaper often published stories about the *Viki Lyne II.* "Why is this boat still in our harbor?" asked one headline on the front page of the *Ladysmith Chemainus Chronicle* above a picture of the rusted-out trawler. In the summer of 2015, about a month before the fires started, an abandoned sailboat had sunk in the Patch, leaving a slick of oil and children's toys bobbing on the surface. The *Viki Lyne II* had thirteen thousand liters of oil in her guts. Over two hundred townspeople showed up at a rally in the Dogpatch held, coincidentally, the morning after the boathouse fire. Over two hundred townspeople showed up. They kayaked through the harbor or stood on Slack Point, waving hand-printed signs begging the federal government to remove the Patch's most notorious vessel.

Occasionally Daniel went on board the *Viki Lyne II* to secure her ropes before a storm—and to shop. He had removed sheets of Plexiglas and spools of rope to use on his boat. The Patch's scourge also doubled as its hardware store for the residents. Pigeon shit splattered her cabin walls, and rust crept up the exposed rebar. Underfoot lay broken glass, empty pop bottles, bent nails, and soft moss that sprouted between the wooden planks. Roosting birds fluttered toward a hole in the roof of the wheelhouse. Daniel once crossed paths with someone carting out a barrel of steel scraps to use for an anchor. If the town moved the *Viki Lyne II,* the people in the Patch would lose a valuable source of building materials. More importantly, they saw the removal as a threat to their way of life.

The relative freedom of maritime laws also unfortunately allows lawbreaking, especially environmental crimes like the dumping of the *Viki Lyne II.* Recycling a boat can be expensive. It's easier and cheaper for an owner to scrape off the serial number and ditch it in the water at night. After the financial crisis of 2008, boats that owners could no longer afford to keep or recycle responsibly began cluttering waterways and coastlines across North America. The

Coast Guard should remove the abandoned boats, but too many derelicts stretch the public purse too far. Unless a wreck is blocking navigation or actively spilling oil, the government directs its attention and budget elsewhere. In 2012, a concerned local first reported the *Viki Lyne II* to the Coast Guard. The Coast Guard commissioned a survey of the *Viki Lyne II* to find out whether or not she was leaking oil. The report found that she contained thousands of liters of oil and her sinking was "imminent." The Coast Guard still declined to remove the trawler.

If the *Viki Lyne II* sank, it could directly affect the fortunes of Ladysmith. Thirteen thousand liters of oil could spread to four nearby oyster farms, to First Nations lands, and to the Ladysmith Maritime Society docks, where people pay to live on their boats—newly renovated and a stark contrast to the Patch's disorder next door. "The liability is tremendous," said Rod Smith, the Ladysmith Maritime Society's marina's managing director. "The suspicion is that the only thing holding the hull together is the marine growth." People in the Patch vociferously denied this. Look at the spray paint on *Viki Lyne*'s hull marking the waterline, they said. The water hadn't moved, so how could she be sinking?

Daniel collected thirty-five signatures from full- and part-time Dogpatchers, as well as a few people who lived at the neighboring marina but hung out in the Patch. Reluctantly, he signed the letter as chairperson of the newly formed Ladysmith Harbour Community and mailed it to town hall. Not long after, Dogpatchers who owned cellphones began to pass him media requests when they crossed paths on the water. He decided it was time to buy a phone.

❧

If there was a sheriff in the Dogpatch, it was Traci Pritchard. She was forty-seven years old, bore a passing resemblance to Pamela Anderson (possibly the most famous person to come from Ladysmith), and had muscular upper arms and a husky laugh. "It was pretty sketchy," she said about her early years in the Patch. "A lot of partying in the evening and drugs. I decided I had had enough when someone broke into my boat. I took care of everything after that."

The marina next to the Patch had installed a floating log barrier between properties that forced Dogpatchers to motor their skiffs farther out into the channel if they wanted to reach the docks. Traci decided to build a dock for the Dogpatch. No one offered to help, and eventually, she abandoned the project on the shoreline. "I would not want to get into a fight with her," said Bryan Livingstone, "because she'd win, hands down." Others were less sanguine. They resented how Traci careened forward, consulting no one.

Splintered relationships like this were common in the Patch. Hardly anyone was universally liked, and alcohol was the most polarizing of vices. The sober Patchers—whether they were recovering alcoholics or longtime abstainers—isolated themselves from the rowdy drinkers. Long-standing feuds had grown out of perceived slights. Some neighbors never spoke, others negotiated on-again, off-again friendships. Everyone had a bit of gossip on everyone else.

When Daniel first arrived, Traci introduced herself and lent him a chain to anchor his boat. These neighborly introductions had a survivalist bent to them: the Patchers wanted to feel out the newcomer, figure out whether he would upset the delicate Dogpatch ecosystem. Traci was one of the few people who would actually step in when someone disrupted the calm. Her efforts had led to nose-to-nose shouting matches, brandished flare guns, restraining orders, cut lines, missing equipment, and whispers about who was responsible for what crime. According to her, she often jeopardized her personal safety to bring order to the Patch. On the inside of her shallow pink skiff she scrawled a warning: "Beware the dogs; they eat everything I shoot."

Living on the water meant giving up the amenities of land. This was the deal everyone in the Patch made with the water. They didn't pay rent or taxes to live there, so they couldn't expect things like hot showers, indoor plumbing, central heating, or emergency services. Neither the town's police nor the firefighters owned a boat, so when fires broke out in the Patch—a common occurrence on old boats with wood stoves—the volunteer firefighters waited onshore until the Coast Guard or the Royal Canadian Marine Search and Rescue boats arrived. By the time they did, the boat was usually burned to the waterline. When the boat next to Traci's float home caught fire, she stayed up all night throwing water on her house. She valued self-reliance. Fewer rules meant more freedom—but also more risk.

It was not easy for Traci to be one of the few women in the male-dominated Dogpatch. When she first arrived, she was only the second female resident. Before earning her spot as the de facto authority figure, she went through a hazing process. "The guys here would pull tricks on me, pull my anchor, and I'd wake up and be on the shore. They ran the other girl outta here, but eight years later, I'm still here." At the time I interviewed Traci, two other women were living in the Patch. One was a recluse who rarely came out of her boat, and the other was a new arrival who declined an interview.

In her nearly fifty years on Vancouver Island, Traci has lived many lives. She was married for eleven years, and she has two adult daughters. "I was very transient after my divorce," she said. She moved from shelter to shelter, struggling with alcohol and drugs and the lingering trauma of a past assault. When she described how a man stalked her for two years before confining and beating her, her eyes locked on some distant point. Traci arrived here in 2009, after she lost her first boat home in Nanaimo, a small city twenty-five kilometers north of Ladysmith. The port authority had instituted insurance fees in the harbor where Traci lived, and the owner of the boat she rented pulled it from the harbor. Suddenly, Traci was homeless.

Traci still wanted to live on the water. "I found it kept me moving," she said. She bought a new boat for a dollar from an owner who had planned to ditch it in the Patch. She christened the boat *Plan B.*

As Traci rebuilt her life and *Plan B*, she lived on next to nothing. "I wore my snowsuit to sleep, had no heat, no nothing," she said of that first year. Eventually, she refurbished and resold *Plan B*, bought another boat, fixed that one up, and resold it for the most cash she'd ever held in her hands: $4,500. That became the down payment on a 2,000-square-foot floating shingled cottage that she called *Anchor Management*. She said she'll die on it.

The name of her home was not just a sly riff on the term *anger management* but also how Traci saw her overseer role in the Patch. She can count eight people she has personally rescued over the years. Once an elderly couple couldn't haul up their anchor and drifted right into *Anchor Management*. Traci opened her front door to see the bow of their boat flush against her dock. She spent an hour pulling the heavy, slippery chain from the depths. A few years ago, she had a heart attack. A wharfinger at the marina heard her calling for help. He drove her to the hospital; in her delirium, she called him Dad.

Living on the water meant living with daily risk but also a higher code of conduct. On land, you might walk past someone passed out on the sidewalk, but on water you stop and help. Article 98 of the United Nations Convention on the Law of the Sea states that wherever possible, every master of a ship has a duty to help someone at sea. Some Dogpatchers took this as a point of pride. To them, living on land bred a lack of responsibility for others and moral failing. "When someone is in trouble, drowning, you help them. It's incumbent upon you. Or when a vessel is in trouble," Bryan Livingstone said. "There are these moral codes that govern the world of boating that not all people up there follow." He pointed a thumb toward town.

❧

Three days before Ladysmith was expected to clear out the boats, Daniel spoke anxiously into his new cellphone. After he'd mailed the letter and petition to town hall, the mayor and Ruth Malli,

Ladysmith's city manager, had invited him to an informal meeting, in which the two sides had agreed to work together. Although the town hadn't guaranteed that it wouldn't remove boats on the deadline, Daniel had left the meeting feeling optimistic. Now he was on the line with Malli. "Can we get a confirmation or a statement that come November 15, no further action will be taken?" he pleaded.

Up and down the west coast of Canada and the United States, cities and towns were rolling back the right to live on the water. Vancouver, the closest big city, had recently rewritten the shipping rules of a downtown inlet where boats had anchored for over a century. A group of boat-dwellers contested the rule in court, but the provincial judge sided with the city. Anchoring for an indefinite amount of time was not a right, the judge decreed in a landmark ruling that has since been cited in a few nearby cases. The boats in downtown Vancouver now have to shuffle along every two to three weeks.

Similar cases have cropped up farther south. In Oakland, California, a counterculture city that is undergoing a wave of gentrification, living on a boat has become a common, even trendy, source of cheap housing. In California, the operators handle life at a marina, and many cap liveaboard slips at 10 percent of their overall stock. As more people opt to live on boats, this leads to long waitlists for spots. Some become "sneakaboards"—people who secretly live at the marina while they wait. Or they choose to anchor on open water until a spot frees up. However, the system can only support so much rule-breaking until it starts to crumble. In nearby Emeryville, one marina had so many sneakaboards that it evicted some twenty boats in 2019. Around the same time, local police demolished ten boats that were anchored out in the Oakland estuary. While the people in the Patch maintained they had a right to live on water and bristled when they were called squatters or trespassers, they knew their claim was tenuous.

Days passed, then a week, and Daniel heard nothing. His calls and emails to Ruth Malli and the mayor went unreturned (Malli said she always called him back). When he went to the town hall, he was told neither of them was there. Then, on the night of November 11, four days before the eviction deadline, the first storm of the season screeched into town. On a boat, there are many noises: creaks, moans, slaps, slams. Eventually, you learn to tune out the everyday din. Around 4:00 A.M., Daniel woke to an unfamiliar bang.

He climbed out of his warm berth and slipped on his flip-flops. Outside, rain flew in sharp horizontal blasts, and waves crashed in from the southeast. Daniel discovered that one of the car tires that protected his boat from rubbing against the cedar dock had been mushed into a flat strip of rubber. One big wave would raise her up and slam her down, and might crack *Sojourn*'s boathouse clean through. In the dark, Traci passed by in a skiff, shining a flashlight and checking boat lines. Waiting for a longer break between the waves, Daniel squeezed a new tire between the dock and his home.

Later that day, Daniel finally got Ruth Malli on the phone. She told him she needed other jurisdictions to approve an extension of the deadline, and she didn't have their go-ahead. The town still expected people to move by November 15, and she asked him if anyone had done that yet. "No," Daniel told her. "People are fearing the ocean more than your eviction." He rubbed his eyes as he sat hunched over the table in *Sojourn*'s galley. He didn't have enough time to find a new place to anchor or to pack up his boathouse—with the generator, smokehouse, skiff, and floating docks—and haul it all away. And what if he left, and the town decided to work with the people who remained? He felt he didn't have a choice but to stay.

"I better not get fucking arrested for this," he said.

Not long before the eviction deadline, Traci dragged a basket of colorful laundry along the marina's docks. Even though the Dogpatch and the marina were uneasy neighbors, many people in the Patch still showered, filled up their water tanks, and used the laundry machines at the marina. Traci was one of them, and she was wearing her last article of clean clothing: bright purple pajamas patterned with bug-eyed, neon-green frogs, tucked into a pair of gaping rubber boots. She loaded the laundry onto her pink skiff, piloted it out and around the marina's breakwater, and hoisted the basket of clothing over the gunnel and onto the dock, accurately calculating the perfect moment to step from swaying boat to swaying dock.

While some Patchers waged a never-ending battle with the wealthier marina next door, Traci had become its most active volunteer. She processed its recycling in a special boathouse crammed with bottles and a wheelbarrow. During the summer, she volunteered as a boat greeter. Each slip had a planter that Traci and other volunteers decorated with fake flowers, old perfume bottles, driftwood, and other finds from beachcombing excursions.

The difficulty of living on water seemed to be good for her, breeding an intense, laserlike focus on the task at hand and shutting out distracting thoughts. "I'm constantly moving here," she said. After spending a week in the Patch, I started to feel it myself. Life on land demanded so little of me physically. Doors swung open automatically, toilets flushed unasked. I could multi-task my way through life without ever looking up from my phone. At sea, routine chores were complicated, composed of multiple, sometimes dangerous, steps.

The night before the eviction deadline, I borrowed Daniel's skiff to visit a sailboat just outside the central cluster of the Dogpatch. As I rowed past the giant white steel breakwaters and out into the main channel, the waves grew lumpy and tall. The sailboat I wanted to visit—a 32-foot cutter bucking in the chop a few hundred meters away—looked farther out than it had on land. A little light blinked in the porthole. The man who lived on board, a retired nurse and recovering alcoholic, was definitely home, but he was nearly deaf and didn't know I was coming.

As the little skiff pitched, I pulled at the oars and tried not to think about how out of my depth I was here on the dark water. If I slipped climbing on board the sailboat or tying the skiff to its railing, it would be hours before anyone thought to look for me. It was mid-November. The water was a degree or two above freezing. I was a strong swimmer, but not strong enough to make it back to shore before hypothermia or worse set in. When I pulled myself on board and made it into the sailboat's cabin, my hands were shaking from the cold and my nerves. The powdered hot chocolate the liveaboard, Ken Ireland, offered me had never tasted sweeter. Ken spent thirty years in Alcoholics Anonymous, and his sobriety, along with his distant anchoring location, placed him slightly outside the central hub of the Patch. That night, as he reflected on the roots of his addiction, he cast it as a symptom of society's distorted values. "I didn't know how not to be that person, because we all learn that we want to get the pretty girl, get the drink, go to the parties, and end up with a girlfriend. In *Playboy*, the girls are all perfect; they're all airbrushed. We grew up with this false sense of reality. You want something, you buy it on credit. Well, all these things led to stress." In the Patch, he found a feeling of peace. He saw it reflected in others who lived here, too. An hour later, I rowed back. Ken kept a searchlight on my dinghy until I made it back inside the calm water beyond the breakwater.

The challenge was what the Patchers liked. Daniel enjoyed the difficulty of personally figuring out what ailed his engine. Traci had a sense of accomplishment from cleaning up the Patch, collecting recycling, and greeting visitors. But she thought sometimes about moving *Anchor Management*, leaving behind the pollution, the infighting, the never-ending scramble to survive. "Living here takes a toll," she said as she showed me a picture of herself when she first arrived. A less weathered version of herself smiled from the pictures, a dimple marking her grin. There was a reason why most people in the Patch were healthy, able-bodied men. The hard life broke the body down and drove even the dichards back to land. Norman Brook had lived at anchor most of his adult life, including two and a half years in the Dogpatch. He was fifty-eight years old but looked much older, with a long white beard, sun-worn skin, and friendly hazel eyes.

Four years earlier, Norm had received an injury settlement from Veterans Affairs Canada, and he and his wife had moved into a slip at the nearby marina, where it cost them $620 a month to park. Cutting up wood for warmth and motoring back and forth to land to walk their two dogs had become too much for them. Ever since he was injured in the navy, Norm suffered from chronic pain and headaches. His wife had fibromyalgia. But still, they visited the Patch often, and if their health had permitted, they would have chosen the Patch over the public dock. "The only real security of person I have ever felt is living at anchor," Norm said.

Hauling water, heating a home with wood: these were antiquated and time-consuming chores. But they were the price people in the Patch were willing to pay for a simpler life at sea. Another reward was the beauty and quiet of living on the water. Most of the year, traffic wasn't an issue—summer was the exception. The tides ebbed and flowed two times a day, pirouetting boats around their anchors. One resident had a trick for figuring out if the tide was rising or falling. Dry land above the waterline—tide coming in; wet land above—tide going out.

Every morning, Daniel sprinkled oat flakes into the water for the minnows beneath *Sojourn*—his aquarium, he said. The resident seal was like a neighborhood dog, following people as they rowed back and forth to their boats. A family of otters moved as one wriggling mass. They were the raccoons of the Patch, crawling into dinghies, eating everything, shitting everywhere. When Traci heard the sweep and flap of a blue heron passing her window, she peeled back the curtain to see which dinghy was approaching her dock. The herons were good watchdogs.

❧

On eviction day, the Dogpatch was at maximum capacity. Every owner was on board and waiting. It was gray and cold, a shiver of rain sprinkling the water surface. As the day progressed, the sun broke weakly through the clouds. A convivial feeling crept up. For once, everyone was in the Dogpatch for the same reason. People who hadn't shown their faces in months were on deck, sipping beers and watching for the town's next move.

Daniel spent most of the day fiddling nervously with *Sojourn*'s rebuilt engine. That's how he used to spend his time before he became an unlikely spokesperson for an unexpected movement. He had given up his time, his privacy, and his plans, and in the end, those sacrifices hadn't protected his boat or anyone else's. If the town came to haul away the Dogpatch boats that day, he wanted to have the engine running so he could make a quick escape and save *Sojourn*. He squirted more oil into the engine and got back to work.

Throughout the day, a stranger went through the Patch dropping off envelopes with the words *Fight the Eviction* scrawled on the front. When he got to Traci Pritchard's dock, she told him off for coming aboard without permission, until she saw the envelope. Inside was a letter claiming that the Dogpatch rightfully belonged to a lessor in Calgary who owned the land bordering the water lot. Without that owner's go-ahead, the letter said, the town had no right to remove boats.

After hours of trial and error, Daniel's motor finally roared clean and loud. He backed *Sojourn* out of his boathouse, keyed up and excited. As he rounded the Ladysmith harbor, however, one of the engines failed. With the remaining engine, he turned *Sojourn* around and took her limping back to the boathouse. A few days earlier, the engine on Daniel's skiff had also died. Until he found a cheap replacement, he would be stuck paddling around the Patch. A trip to the marina had once taken five minutes; now it took twenty.

Later that evening, Daniel poled his skiff gondolier-style over to Vince Huard's two strung-together boats. A small party was gathering. Vince had hauled up Dungeness crabs from farther out in the channel, and now he was at the barbecue, slathering them in sauce.

As Vince grilled crabs on the deck, Daniel took a seat in the wooden galley. Across from him, Luke English ate a bowl of spaghetti and flipped through a book of knots. Everyone called this 26-year-old, long-haired, tree-planting troubadour Newbie. He had arrived three months ago and knew nothing about boats, but the Patchers liked him all right so far. He set aside the empty bowl and strummed a few chords on his guitar. "I'm a happy noodle man," he sang to himself. "I'm Italian."

Outside, the sky faded from blue to pink to black. Vince flicked on his headlamp to watch the crabs turn from gray to red on the barbecue. "You stay as long as you can, and then they will kick you out. Why? 'Cause it's the best place on the water," Vince said. "We don't have rights. It's all at the whim of when the water lot is sold to someone. Then we gotta go. Meanwhile, we're gypsies. We gather in an area, hang out together, protect each other, and then eventually we all scatter and find a new place."

Packing up a subsistence life would be more challenging for the older, the poorer, and the disabled in the Dogpatch. If the town removed the three strung-together boats that an eighty-something deaf mariner lived on, where would he go? Or the two hoarders who got by on disability checks? Or Lew McArel, whose monthly $244 welfare check would barely cover food, let alone accommodations, on land. "Go where?" Lew asked. "I'll just live on the streets." The townspeople of Ladysmith had legitimate concerns about the Patchers: the way they heated their boats with shoddy wood stoves and their bucket-and-chuck-it approach to sewage. But in trying to deal with abandoned, derelict vessels like the *Viki Lyne II*, which might or might not have been sinking into the sea, the town threatened the lives of a group of mismatched outcasts who had slipped through the cracks on land and found, if only for a brief time, home.

As cities build a more connected relationship with their waterfronts, will we accommodate people who were living on the ocean before the mainstream rediscovered it? Will we look deeper at their reasons for being on the water? Or will we impose new rules that strip them of a space, and call it progress? The party inside Vince's boat grew louder, voices and guitar strums ricocheting off the tight interior of the wheelhouse. Outside, the sky darkened and the stars were reflected in the ocean below, creating a 360-degree spin of bright white pinpricks in the jet-black sky. The conversation swung to the eviction deadline. No town official had shown up. Either the Dogpatch had called the town's bluff, or they were living on borrowed time.

"We're here, aren't we?" Daniel said, raising his glass for a toast. Newbie improvised a new song on the guitar. "They want to kick you out of town," he sang in a bluesy twang. "But at least your boat hasn't burned down."

FIVE

CLEANING THE COAST

A worn piece of plastic drifted on the ocean over a thousand miles from civilization. A sailboat approached with a thirty-year-old woman on board. She leaned out over the gunwale to pick the plastic from the surface. Except she couldn't: long, dangling seaweed roped the plastic to the water. She reeled up the weed, hand over hand; it stretched deeper and deeper into the depths. Down below, she saw fish darting between the fronds.

As Chloé Dubois sailed farther into a slowly spinning gyre of plastic in the largest ocean on Earth, she experienced this scene again and again. It was 2015, and Chloé and her team at the nonprofit Ocean Legacy had sailed to the Great Pacific Garbage Patch to collect microplastic samples for The Ocean Cleanup, another plastic-pollution nonprofit. Using samples collected by thirty-seven boats, Chloé's included, that trawled a 3.5-million-square-kilometer swath of the Pacific, Ocean Cleanup hoped to create the first high-resolution map of ocean plastic. Chloé remembers hauling up the water-sampling trawler and peeking in at its contents on deck, and discovering all manner of marine stowaways in the detritus. *How did you get here?* she wondered as she picked up a tiny crab clinging to a bottle cap in the middle of the formidable ocean. Drifting by the boat, she saw buoys covered with gooseneck barnacles. Ocean-knotted islands of rope that hid masses of organisms.

"On the news, there's this plastic island in the middle of the ocean that's the size of Texas, and that's pretty much what people know unless they go out there and experience it for themselves," she said. Instead of a floating island of waste, as the Great Pacific Garbage Patch is so often portrayed, she encountered more of a drifting slurry. The pollution came in all shapes and stages of degradation, from microscopic particles and fibers, to toothbrushes, bottles and great tangles of fishing nets and lines. She witnessed, too, how nature worked with the plastic intruders. In the ocean, bacteria and algae quickly glom onto any floating feature they can find, drawn to the nutrients that collect there. More and larger animals, like barnacles and tubeworms, follow suit, fastening themselves to the marine debris. How productive of the ocean to use the plastic to build tiny ecosystems out on a vast desert of salt water, where so little life thrives in comparison to coastal waters. The Garbage Patch was not a dead zone at all, she realized, but a world teeming with life.

Since she was seventeen years old, Chloé has been involved in the environmental movement. In her early twenties, she began collecting plastic from beaches and she's now cleaned shorelines across Mexico, Alaska, Costa Rica, Panama, and Canada. When she was twenty-nine, she cofounded the nonprofit Ocean Legacy, and she has become obsessed with cleaning plastic from the environment. She knows the names, acronyms, and resin codes of the plastic pantheon like they're her children.

For a moment, Chloé hesitated before destroying the little crab's home, this plastic piece of garbage that it had found and colonized and survived on against all the odds. Rationally, she knew that the crab's plastic bottle cap was on its way to becoming a toxic pill. Plastic is a master at teasing out toxins from the ocean, sucking floating chemicals from the water column and condensing them into ever more hazardous forms. Industrial metals, pesticides, fertilizers, plastic softeners, and flame retardants can dissolve in water or be hydrophobic, meaning they want out of the water fast. Plastic already contains some of the chemical contaminants found in water, and that makes certain types of plastic naturally attractive hosts to wayward chemicals. A smaller animal might then ingest that poisoned plastic item, covered in slimy nutrients and pollutants, like PCBs, that have been banned on land for decades but are still drifting out in the ocean. A larger animal will then eat this animal, and up the food chain the plastic goes, magnifying its toxicity as it jumps to each new animal. Chloé knew all this. She had seen the damage firsthand, yet destroying an animal's home still gave her pause. Then she plunged her hands in and removed all the plastic she could find, no matter how much life clung to it. The team built a home for displaced crabs in a glass tank on deck.

When they had sailed outside the center of the Great Pacific Garbage Patch, Chloé dove off the boat and into the sea. When she climbed back on board, tiny pellets of plastic covered her skin. After a month and a half sailing across the Pacific, her sailboat returned to land with 154 water samples hauled up from across the ocean. Every single one contained plastic.

Not all plastic is a problem. Much of it helps us and is integrated into every step of human life from birth to death. As I write this, I tap away on computer keys made of plastic, scroll through webpages on a mouse made of plastic, and peer through glasses rimmed with plastic. It's the cheap, omnipresent plastic that lasts hundreds of years but is built to throw away the second after we use it that's a big problem, perhaps one of the biggest for the ocean.

For almost as long as industrial plastic production has existed, we've known that plastic was going in the ocean. In the 1970s, a team of researchers sampling water in the sluggish Sargasso Sea reported that tiny plastic fragments were floating on the surface. During a 1997 yacht race from Los Angeles to Honolulu, a sailing scientist named Charles Moore passed through a remote stretch of Pacific Ocean and found himself surrounded by plastic debris in all directions. The Great Pacific Garbage Patch, as it was later called, grabbed the world's attention. Suddenly there was a tangible place where all our waste was going, just outside the limits of our imagination.

The sea is a vast, deep, mutable force that covers 71 percent of the Earth. Plastic is small, ubiquitous, and breaks into ever-smaller pieces. When these two meet, they marry: a horrible collision between the synthetic and the natural. Given enough time, the ocean has the ability to spread plastic to the most remote reaches of the planet. Today, plastic is drifting in the waters off Antarctica. Plastic comes down in rain. Plastic fibers pass through the filter-feeding valves of oysters. Not long ago, Japan's Deep-sea Debris Database reported finding a fully intact plastic bag in the Mariana Trench, the deepest underwater trough in the world.

We still don't know exactly how much plastic is going into the ocean. One study, published in *Science* in February 2015, conservatively estimates that eight million metric tons of plastic is entering the ocean each year from municipal solid waste streams on land. That is two hundred times higher than what had last been calculated in 1975 based on plastic pollution entering the ocean from maritime activities, and more than two thousand times higher than what had been estimated from floating debris samples. In that 2015 study in *Science*, environmental engineer Jenna Jemback and her coauthors argue that barring any major changes, plastic going into the ocean will multiply by a factor of ten in 2025. That's eighty million metric tons of plastic dumped in the ocean each year. Despite the startling numbers of waste already in the ocean, our love of plastic endures. Plastic production is growing and expanding right along with plastic demand. By 2030, our need for plastic is expected to double.

The financial guru Warren Buffett once compared a stock market crash to the tide going out: you find out who's been swimming naked all along. During the 2008 financial crisis, we discovered that big banks can fail. For centuries, we've believed the same of the ocean: that it was simply too big to fail. But an encroaching movement of threats, such as a warming ocean, overfishing, and pollution, could change that in the not-too-distant future.

If we could see beneath the surface, what would we find at the bottom of the sea? Perhaps millions of tons of plastic lying undisturbed, except for the bottom-dwellers that nibble at the nutrients collecting on it. Perhaps this evidence of the world's waste will eventually become a layer of sediment pressed between rock layers: the plastic era, a fitting symbol of human-made change, baked into the Earth's crust.

"Goooood morning!" A singsong voice echoed across the water to a collection of tents on the shore. A moment later, French-Canadian hip hop, blasted up to eleven, boomed out of the fifty-foot trawler where Chloé Dubois had slept the night before. This was how she woke her volunteers each morning for a long day of collecting plastic along the jungle shorelines of the Pacific Northwest. In these early morning moments, as I surfaced in and out of sleep and Chloé's painfully cheery good-morning drifted in through the thin walls of my tent, I could see how she earned her nickname: Mary Poppins.

A year had passed since Chloé's trip to the Great Pacific Garbage Patch and The Ocean Cleanup, an organization that plans to utilize wave power to clean the ocean surface, was analyzing the samples she'd collected. In the meantime, her work with Ocean Legacy continued, organizing hard-core plastic cleanups in remote territories. This was their most hard-core trip yet. In 2013, a local told Ocean Legacy about MuQwin Peninsula and the obscene amounts of plastic that collected on this cape jutting out into the Pacific Ocean. This was her third summer here with a team of volunteers to collect and remove plastic. By the end of the six weeks, Ocean Legacy would haul out twenty-five metric tons of plastic from MuQwin on a barge they had also salvaged from the shoreline.

With permission from the local Checleseht chief, they set up camp on a small island and used machetes to clear paths and campsites out of the dense rain forest. Early on during their stay, the team startled a black bear that swam for the mainland. While cleaning up just this island, they collected twelve giant sacks of ocean plastic. This might sound like a small, insignificant action in the global context of plastic pollution, but when plastic washes ashore, this is our chance to stop the spin cycle by removing it from the environment forever. Otherwise that plastic washes back out to sea and becomes nearly impossible to extract from nature.

The day before, Leo, the water taxi driver, had dropped me and a French family spending their summer volunteering with Ocean Legacy at a local dock in the channel at Kyuquot. A fifty-foot trawler waited to take us the rest of the way into MuQwin. On one side of the channel was a restaurant with Wi-Fi and a public telephone. Leo ran a grocery store on the other side; ice cream was in stock. Behind the houses hanging over the ocean, paths linked the homes and ran into the forest. Water was the quickest way to get around. On board the trawler we met the plastic-collecting team we would be working with in the coming week: a former circus strong man, a mechanic with fifteen bolts holding his face together, and a chef who collected oily whalebones by her tent. The core Ocean Legacy team consisted of cheery blue-eyed executive director Chloé; her partner, James, a heavy-set assertive type who captained the ship; and thin, long-haired Dexter, the ship mechanic. There was also Bob the chemist, with a desk job in the Alberta oil fields and perhaps the most straight-edged of the bunch; a recently divorced firefighter who couldn't stop talking about his son; pigtailed Gina, who made up songs about the garbage she collected; and a half dozen other longtime volunteers and recent additions. These sunburned vagabonds had spent weeks collecting and hauling plastic. Everybody had different reasons for volunteering with Ocean Legacy, Chloé said, "but we attract a lot of people who want to do good in the world. It's the unknowing—that the ocean itself is mysterious, romantic—and this idea of what's out there."

That morning, our team slowly assembled, stumbling out of tents in gumboots and walking up bleary-eyed to the campfire. Pancakes fried in a cast-iron pan; people slugged coffee out of metal mugs. Two nights earlier the team found themselves stranded on a nearby beach after the tide went out, shivering in the darkness and waiting for the tide to rise. Everyone yearned for batteries, lighters, cigarettes, and chocolate. The island offered no cell reception, no showers, and no laundry. Instead, people rubbed and slapped their clothes on the rocks of a freshwater creek. Through sizzling sun and pounding rain, they had been collecting and hauling plastic twelve hours a day for twenty-one days straight. They chugged C4, a high-powered energy drink that Dexter got everyone hooked on. Halfway through the six-week expedition, they looked deeply sunburned and exhausted.

By 8:30 A.M., we cleared out of camp and boarded the trawler, heading for the cleanup site of the day. "Today we're looking for the elusive marine debris," Dexter narrated in the Aussie accent of Steve Irwin, the crocodile hunter, as we drew closer to shore. We loaded into the dinghy and headed to land.

The beach was a long stretch of white sand backed by green jungle and high cliffs. Apart from the steady boom and hiss of waves, it was quiet, deserted, and picturesque, the kind of empty beach paradise that exists only on postcards. As our team trudged up a dune to drop the day's supplies, we found a shaded tiki bar and benches and tables under a grove of trees. Farther along the beach lay a raft constructed from fishing buoys and clear plastic sails. A troop of Boy Scouts had cleaned this beach a few weeks ago and left behind a few markers of their visit.

As I walked the beach, my eye sharpened to the straight lines and bright colors of plastic mixed in with the ragged shapes and dusky colors of the shoreline. I picked up plastic bottles full of fluid, which I opened as far from my nose as possible. Strange brews of algae and seaweed came splashing out onto the sand: gag-worthy concoctions of chemicals and shit and a deep bottom-of-the-sea stench. In one I found a soup of carrots and peas. Another was full of engine oil. I stopped pouring it on the sand as soon as I realized.

Chloé set up three giant white sacks side by side on the sand: one for foam, one for plastic bottles, and one for mixed plastics. These industrial sacks were built to hold up to a thousand kilograms. When the tallest sacks were stretched to their full height, they loomed over Chloé's head (she is five foot four). As we collected, these sacks overflowed with plastic.

I scooped plastic bottles and mangled foam into my garbage bag, and it seemed to me that the pollution was not as devastating as I had thought. I expected to be staggered by the sight of ocean plastic. Online, I had seen pictures of remote Polynesian atolls covered in plastic bottles, and I'd assumed I'd see drifting dunes of foam pellets or would break down crying at the sight of branches tangled with detergent bottles like some apocalyptic Christmas tree. It was not like that; it was more diffuse, more diluted by nature. Then again, the Boy Scouts had cleaned this beach just a few weeks earlier, and more than enough trash had collected in the intervening weeks to busy eight volunteers.

The mangled plastics I couldn't recognize I handed to Chloé, who said in an instant whether the plastic was some piece of fishing equipment or a shoe sole or a coat hanger. If she was unsure which plastic resin it was, she sniffed the material or bent it in half. James, the more cavalier of the two, whipped out a lighter and watched how it burned. Sometimes the ocean had so thoroughly warped a piece of plastic that a burn, bend, or smell test wouldn't reveal the precise polymers contained within.

The basic chemical structure of most plastics is a long repeating chain of carbon, called a polymer. The building blocks of most plastics are hydrocarbons, a compound of carbon and hydrogen, which are manufactured from natural gas and coal. As the oil and gas industry boomed during the 1950s and '60s, plastic production boomed as well, converting the leftover hydrocarbon offcuts into plastic. Ever since, the production of plastics has intertwined with the fossil fuel industry, growing and expanding together, particularly in the production of single-use packaging that poses such a threat to the ocean.

Carbon is the backbone of plastic because it plays well with other elements and molecules. Throw in heat and pressure, as well as additives like coloring, softeners and flame retardants, and plastic can become whatever we want. This ability to chemically tailor plastic gives plastic its immense variability and massive diversity of shapes and textures, colors and purposes, from nylons and fleece sweaters to aviation parts and computer keys. It can bond tightly or loosely with oxygen, nitrogen, fluorine, styrene, ethylene, phenol, and formaldehyde. The structure of the polymer chain can be weak or strong, double or single-bonded, loosely or rigidly organized. The structure of that chain also dictates whether a plastic will be hard and durable, like your refrigerator door handle, or flimsy and stretchable, like a plastic bag.

The irony of collecting plastic in plastic garbage bags that were stuffed into even bigger woven plastic sacks wasn't lost on me. I asked Chloé about using plastic to remove plastic. "For the work we're doing, it's pretty intense conditions," she said as we took a break on the sand. "Plastic still holds its buoyancy and is able to withstand the elements. If there were an alternative that was just as durable as plastic, we would use it." Even when it came to collecting plastic, the best tool for the job was plastic.

Chloé was a realist. She didn't believe that the modern world would rid itself of plastic. It's too useful, too rooted in daily life, habits, culture, and industry. Even today, when more people than ever know about the dangers of cheap, disposable plastic, we can't swear off it completely. We could stop using single-use plastic bottles and bags tomorrow, but many high-tech industries, like manufacturers of heating and cooling systems, cars, planes, electronics as well as the health care industry and scientific research, will always need virgin plastic for fail-safe, uncontaminated parts. Our world order depends on producing more new plastic.

Using plastic didn't conflict with Ocean Legacy's vision the way it might with other zero-waste organizations. Chloé didn't want to eliminate plastic; she wanted to change how we thought about and managed it. A year earlier, Ocean Legacy had purchased a one-kilo plastic-to-fuel machine from a California company called Resynergi. The machine worked like a whiskey still but for plastic, vaporizing debris in an oxygen-free chamber, condensing the mixture through cold water, and creating a honey-colored light crude oil. The one-kilo transformation took around four hours, all powered by low-voltage electricity. At the end, foam was reduced into styrene oil, a colorless liquid that could sell for three dollars a liter at the time, according to Ocean Legacy. Plastics like polyethylene and polypropylene were converted into diesel fuel that could power cars or boats. The machine was also an educational tool: small enough to pack up and take to fairs or schools and demonstrate another fate for ocean plastic.

By giving something disposable a value, Chloé hoped that people would treat plastic like a commodity. This is a huge mental shift for the world to make. But she wasn't alone in pushing for a capitalist solution to pollution. A growing movement of entrepreneurs and inventors is trying to assign value to something most people consider worthless. Several building companies are experimenting with creating bricks out of recycled plastic shards. Fashion companies are spinning ocean debris into thread to sew stretch pants and bathing suits. Another innovation melts down empty plastic bottles for use in 3-D printers.

Chloé had a use in mind for her liquidized plastic fuel: she wanted to sell it on the open market. She dreamed of buying a one-ton plastic-to-fuel machine and towing it up and down the west coast to small towns like Kyuquot, whose waste services could not cope with an ocean's worth of plastic washing in. She would vaporize plastic into fuel on the spot and sell it back to the community. But she and the team had to keep digging into their own pockets to make the next project happen. They couldn't go on like that forever.

We sat in the shade of the plastic-stuffed sacks, and I listened as Chloé's chipper voice fell flat and her smile slackened into a straight line. A realist to the end, Chloé knew that her dream depended on money. Recently, a part on their ship, the *Imperial*, broke and stalled the operation for a few days. "James and I have been covering additional expenses," she told me. If they're going to continue growing, Ocean Legacy has to make the financial leap from nonprofit to a business model that will fund their next big dream. But the world is still not willing to pay top dollar for recycled plastic or its liquidized forms, even if it makes environmental sense.

Virgin plastic is easier and cheaper to source than sending people into the bush for six weeks to retrieve twenty-five metric tons of brittle, degraded plastic. When the price of oil drops, virgin plastic gets even cheaper to produce. They haven't been able to sell their fuel or liquidized plastic oil—yet. For now James took the concoction they churned out of the demo machine and dumped it down the *Imperial*'s thirsty gullet. At the end of a long day of cleaning, James revved the engine and the team chugged off to the next shoreline cleanup, a day's worth of plastic lashed to the deck.

❧

On the second morning of the trip, the *Imperial* dropped us at a beach that Ocean Legacy had never cleaned before. It seemed much like the first beach: an underwhelming amount of garbage, as well as a forest so dense and thorny that we couldn't claw inside to clean. Instead, we trudged along the periphery, plucking bottles and buoys from the driftwood and branches. Plastic doesn't land neatly on the waterline. It is blown back into the tangled forest, where it gets caught on brambles or jammed between driftwood. The heavy hiking boots I wore seemed preposterous in the heat, but they were the only protection I had against wobbly driftwood that would suddenly roll and crush my toes. To ease the slog, we plastic-pickers began a competition called Who Found the Weirdest Thing? The entries that day were a motorcycle helmet, a lithium battery with scary stickers asking that we return it to the military, and a toy dinosaur.

Form and function dictated what we found washed up on the sand. Over fifty percent of what Ocean Legacy collects from beaches along the west coast of North and Central America is foam, a puffed plastic made mostly of air and polystyrene—a liquid hydrocarbon that is likely carcinogenic. A lightweight, bulky piece of foam sits top heavy above the waterline and has a greater sail area that the wind can push to shore. Plastic bottles were one of the next major finds. Made of polyethylene terephthalate (PET), a plastic bottle is denser than seawater and sinks as soon as it's punctured. In Ocean Legacy's experience, people are strangely good at screwing caps back on, so more bottles float to land intact than you might expect. What we were collecting on the beach was the lightweight cream of the crop: straws, plastic bags, cups, bottles. Cities around the world are drafting plastic-bag bans and introducing reusable-cup trading schemes for many of these single-use items, and that's what is visibly washing onshore or floating at the ocean's surface. It made me wonder what else we might ban or regulate if we could see all the heavier plastics and materials that never make it to land, simply because of their composition.

"My guess is that if something is sitting at the bottom of the ocean, it will be there centuries from now," said Kara Lavender Law, an oceanographer at the Sea Education Association in Massachusetts. In 2010, Law and her colleagues published a study in *Science* that showed the North Atlantic had a huge circulating gyre of plastic, just like the North Pacific. Analyzing over twenty years of water samples collected from a research boat cruising across the Atlantic and Caribbean Sea allowed Law to demonstrate just how much plastic was circulating at sea. Over 60 percent of the six thousand samples had plastic in them. As many as 167,000 plastic fragments swirled in just one square kilometer of water.

Temperature plays the biggest role in breaking down plastic, Law explained. Little oxygen or light penetrates cold underwater trenches, so the ocean will actually preserve the plastic indefinitely. At sea, plastic has no hard surfaces to bounce off. But when it reaches land, plastic shatters on hard rocks or shreds across exfoliating sand, breaking into tinier and tinier fragments. The sun warms these small pieces, making them more brittle until they break into even smaller chips and chunks. This is when the toxins inside plastic have a higher chance of leaching out into the environment. Studies show that toxins from degraded foam are one to four orders of magnitude higher in the sand than in the surrounding water.

There was no clear winner in our Who Found the Weirdest Thing? competition, but Dexter, the ship's mechanic, claimed the lithium battery to use in his experiments down below in the hull. I pocketed the toy dinosaur. When I had found it earlier that morning, it was starting to degrade, lying on its side, warm from the sun. The ocean had smoothed and worn down its edges. Rocks and sand had crosshatched its skin with faint scratches. It was missing a hind leg. On one side the dinosaur was dark gray, while the sun had bleached its opposite flank white.

There was a small chance that the dinosaur belonged to a victim of the twenty-meter-high tsunami that struck the coast of Tōhoku in northeast Japan in March 2011 after a massive undersea earthquake. The retreating wave sucked five million metric tons of debris into the water. To put this in context: more than half the annual amount of plastic debris that ends up in the ocean went into the Pacific in a single day. Over time, 70 percent of this debris became waterlogged and sank. Whole houses, docks, timber, and bodies settled on the bottom of the ocean. Roughly 1.5 million metric tons floated out to sea. Ships reported seeing battered abandoned fishing boats and houses drifting far from land.

A year passed, and that debris started to wash up along the west coast of North America: a football, a volleyball, then an entire Harley Davidson motorcycle, still inside a Styrofoam container. Miraculously, the owner of the motorcycle was found using the rusted license plate, and the Harley was shipped back to Japan. In the years since, this disaster has offered many opportunities for connection across oceans and cultures. More lost boats have washed up onshore. North American beachcombers have travelled to Japan to meet the people who had lost everything but these prized possessions. The Japan Love Project posts online pictures of found items in an attempt to return them to their owners. Carrying the toy dinosaur with me on the beaches of MuQwin, I thought Japan Love might reunite the dinosaur and its owner. The west coast is uniquely positioned for a devastating earthquake and tsunami of its own sometime in the near future. Each item that washed up from Japan was a reminder of what the west coast could also lose to the sea.

Chloé cringed whenever someone referred to the debris as garbage, junk, or trash. Some of those items probably once belonged to someone who had experienced a devastating loss, perhaps even a sudden and tragic death. Ocean Legacy estimated that a third of what it collected came from Japan's earthquake; a Japanese grant actually funded their expedition to MuQwin. It was hard to know for sure: to qualify as official tsunami debris, the material has to come with something traceable, like serial numbers or place names—a tall order for something that has spent months or years at sea. "An almost judicial level of evidence," as one volunteer put it.

To date, only a small portion of what Ocean Legacy has collected classifies as "official Japanese tsunami debris." Even though there was no way to prove an item's origin without a telltale name or serial number, it was clear that many items came from the 2011 disaster, like hot-water bottles and fishing pallets and flip-flops with Japanese markings. The sheer quantity of Japanese packaging collected was way higher than normal. Yes, this weathered bottle of Skal, a carbonated milk drink popular in Japan, *could* have come from a specialty supermarket, but it seemed unlikely that every Asian supermarket on the west coast was suddenly dumping its garbage in MuQwin Peninsula. This was years after the tsunami, and Chloé Dubois felt that minimizing the amount of tsunami debris also minimized the larger issue of ocean plastic. The debris numbers stayed low, so the story gained no traction with the media. This fit a larger narrative that developing countries far away caused plastic pollution. The 2015 landmark study published in *Science* showed that China, Indonesia, and a few other Southeast Asian countries are responsible for releasing the vast majority of plastic waste into the ocean. The United States, with its long coastline population and high plastic consumption, was the only developed country to make a list in that study of the top twenty plastic polluters, and it came in last place. Globally, our plastic is interconnected. Until recently, developed countries shipped most of their waste to China for recycling. To criticize one country for its plastic pollution while we send our waste to them to manage shows how slanted the story around waste has become.

Every nation is in the middle of its own evolving relationship with waste. North American and European countries set up recycling programs in the 1980s and 1990s. Developing nations are only now starting to reach Western levels of plastic demand. It seems unfair to impose on emerging markets environmental laws that the West never had to contend with when they industrialized. And yet, the ocean's fate and our future depend on cooperation. The ocean is like a superhighway for plastic debris, and every nation with a shoreline has an on-ramp. Regardless of where it comes from, once plastic enters the ocean, it's everyone's problem.

Out at sea, cross currents meet and spin and slowly draw plastic into gyres. Travelling on currents, plastic can work its way from the east coast of the United States to the Arctic. This great mixing of cool and warm water facilitates the journey as plastic flows freely through the sea. Some ocean travellers report coming across great chains of plastic drifting on the surface; others see nothing at all. Ocean plastic can be so minute that it hangs suspended in the water column and only becomes visible under a microscope. How do we deal with a problem that can seem invisible or changes by the day or the hour? That shiftiness gives the impression that ocean plastic is not such a serious problem after all.

Some also dismiss the severity of the problem because they still think of plastic as chemically inert. We drink out of plastic cups, heat up food in plastic, and we store all kinds of chemicals in plastic. If plastic still retains its shape under heat and doesn't dissolve when it touches other chemicals, it must not be harmful in the ocean, right? When animals eat small pieces of plastic, won't they just harmlessly poop out the plastic and continue on their way? This posits that ocean plastic is mainly a litter issue rather than a threat to the ecosystem. What this doesn't take into account is that plastic is designed for a specific purpose, and a possible afterlife in the ocean is not part of that design. When discarded plastic fishing nets drown sea turtles or plastic waste fills the bellies of whales, we're only seeing the early impacts. This is the first stage of figuring out all the unfathomable ways plastic might harm the ecosystem.

Labs around the world are studying the giant, uncontrolled plastic experiment we've unleashed on the ocean. In the environment and over time, plastic starts to weaken, and its chainlike bonds break apart, releasing hazardous additives. These are things like flame retardants, softeners, stabilizers, and thousands of other additives that help plastic items fulfill their purpose. On land, plastics and their additives are mostly nontoxic, except when we find out they're not. One additive, bisphenol A, has become well known to new parents, who now buy baby bottles free of the estrogen-mimicking industrial chemical. Bisphenol A also has links to cardiovascular diseases and type 2 diabetes.

Persistent organic pollutants like PCBs, which have been banned for years, are still out in the water column, attaching themselves to floating bits of plastic. PCBs and other chemicals latch on to ocean plastic. When animals eat the plastic because it's covered in delicious nutrients, the plastic works its way into the food chain. Eventually, that will come back to humans. One study of seventeen salt brands from eight countries concluded that humans eat thirty-seven pieces of man-made particles each year from salt, most of them as big as a coarse grain of sand, while a 2019 report from the University of Newcastle calculated that the average person ate a credit-card-sized amount of plastic each day, mainly through drinking water. Another study conducted on seafood markets in California and Makassar, Indonesia, found that more than a quarter of the fish sampled contained man-made debris, mostly small plastic fragments or unidentifiable fibers. Recently, researchers have revealed that plastic damages coral reefs by blocking out sunlight, releasing toxins, and raising the likelihood of infection as much as twenty-fold.

The possible impacts are virtually limitless. "Plastic knows no borders," Chloé said, and all that plastic was becoming as free as the high seas.

❧

By noon on the second day of plastic collecting, my progress slowed to a stumble. The sun was relentless, and every patch of exposed skin felt like it had been aged and dry-rubbed like a prime piece of meat. I had a pounding dehydration headache. The water I chugged made no difference. My hiking boots sank in the sand like stones in water. I mostly moved from momentum, half-falling through the humid air.

I tried to find a clearing in the woods so I could collect plastic in the shade for a while. I couldn't find a way into the brush. It was a dense wall of branches and foliage, a secret garden locked to intruders like me. I had seen other people climb in here. How did they do it? When someone found a buoy in the forest, a warning call of "BOO-EEE!" echoed from the bushes, and out flew a buoy that landed on the sand with a satisfying plop. This became a rallying call of sorts, the only way to break the hot monotony of walking and stooping and bagging more plastic.

Each Ocean Legacy volunteer had a penchant for a certain type of found object. The freckly chef Mackenzie held on to animal bones, feathers, and a calligraphy pen that still wrote wispy black letters after a long dip in the sea. Pigtailed Gina held on to a rusty chain for her garden back home. Only a fool would toss away the Japanese glass buoys, a rare and precious find in the wild. Chloé had a few glass buoys hanging in her living room. Mackenzie told me about a bottle she found with a note inside pleading for a lesbian lover, phone number attached.

After a stretch of beach was cleaned of plastic, the pile of buoys from the forest waited on the sand until Chloé returned with a frayed fishing line and threaded them on one by one. Then she dragged the giant necklace down to the plastic sacks, and at the end of the day, we hauled everything out to the *Imperial*. Every step of the cleanup was hard, manual labor; nothing about it was easy or automated.

It raised the question: wasn't there a more efficient way to clean up ocean plastic?

A young Dutch inventor named Boyan Slat attracted mass media attention and celebrity investment for an ocean-cleaning invention. Slat proposed a system of free-floating booms that passively strained plastic from the sea using ocean currents. It was an impressive, seemingly logical system—just let the ocean do the work. A TEDx video of Slat explaining the system went viral in 2013. Slat dropped out of his aerospace engineering degree program and started to work on his nonprofit, The Ocean Cleanup, crowdfunding over $2 million and raising more than $30 million from Silicon Valley investors. Environmental A-listers like Leonardo DiCaprio supported the project, too. Chloé got the chance to visit the greatest ocean dump on Earth when The Ocean Cleanup gave funds to environmental groups to go to the Great Pacific Garbage Patch and collect water samples. The Ocean Cleanup would later analyze those samples to learn where to place the ocean-collecting booms at sea and gather the most plastic. Before the technology was tested at sea, Slat won the UN's highest environmental honor, as well as other numerous awards for his commitment to ridding the ocean of plastic.

Unfortunately, few experts believed that The Ocean Cleanup's system would work. In 2013, a group of leading marine debris researchers outlined a litany of reasons, scientific, legal and financial, against cleaning the Great Pacific Garbage Patch with a system like Slat's. The most compelling reasons were logistical. The North Pacific's plastic gyre is huge and has a surface area more than double the size of the United States. It is deep, with an average depth of two-and-half miles. And the plastic found in subtropical gyres is not neatly condensed for easy collecting, with as little as one five-millimeter piece of plastic for every square meter. Transiting such a vast territory to collect even a kilogram of plastic would be an epic and expensive undertaking. The letter went on to describe how an ocean-cleaning system, designed to help marine animals, could actually end up hurting or killing them. A variety of ocean creatures, from whales to zooplankton, could die when they become entangled in nets. A year later, Boyan Slat released a 528-page feasibility report that failed to convince much of the scientific community.

Law and Chelsea Rochman, a marine biologist at the University of Toronto, both support Ocean Legacy's approach instead: remove plastic when it washes up on land, and stop plastic from entering the ocean in the first place. In 2017, The Ocean Cleanup launched a trial boom in the North Sea to much excited anticipation and scientific skepticism. It broke apart, adding presumably more plastic to the ocean. In early 2019, the nonprofit tried again, launching a larger, two-thousand-foot boom off San Francisco that was meant to collect plastic from the Great Pacific Garbage Patch. It also failed to collect plastic and eventually broke down. The Ocean Cleanup has tried to frame these failures as an iterative learning process, but four years earlier, oceanographers had predicted the exact issues Boyan Slat's plastic-collecting boom later encountered at sea.

How did Boyan Slat and his project lure in so much investment? Perhaps we want to believe that technology is the answer. Technology gave us this supermaterial that can be molded and blown and spun into thousands of helpful tools. Why couldn't another technological advancement solve the problems our love affair with plastic has created? Of course, realistic solutions are available right now, like collecting plastic from the shoreline, taking reusable mugs and bags wherever you go, and pressuring governments and businesses to reduce single-use packaging. But those solutions are harder in the sense that they are less convenient for us. The Ocean Cleanup wouldn't address the source of the problem, and it never communicated the irreparable damage we've already done. More than five trillion pieces of plastic, and counting, are scattered across 71 percent of the planet, hiding inside marine animals and buried in deep-sea trenches.

When it comes to solving ocean plastic, we're still in the triage phase. Right now it's most important to stop the bleeding. Walking the MuQwin Peninsula, stuffing plastic in a bag, and hauling it out by hand feels decidedly unsexy and labor-intensive compared to deploying a boom in the middle of the ocean. Unfortunately, that's the reality today. We have no one-size-fits-all solution for the sea.

Later that afternoon, Mackenzie called out to me. "This way," she said, leading me to the edge of the beach and into the shade of the rain forest. I followed her up a towering pile of driftwood, and we dropped down into a quiet, lush swamp bordered by waist-high reeds. The boom and hiss of the surf receded; the dazzling sunshine was gone. There was only the quiet, green shade I had sought in my dehydrated haze—and it was absolutely covered in plastic.

Here was the destruction I had imagined finding when I first arrived in MuQwin. I was standing in nature's cathedral, beautiful sunlight filtering down through a green canopy of leaves overhead, surrounded by ferns and sala berry bushes and trees covered in electric-green moss. Plastic carpeted the ground. Plastic bottles, plastic buoys, Styrofoam everywhere, like someone had Photoshopped a garbage dump onto the forest.

I no longer needed to scrounge for plastic. I squatted, stretched out my hands, and raked plastic into my arms. That was only the first layer. When I moved aside a log, I found yet another layer of plastic and then another, all of it slowly disintegrating into the morass beneath. The swamp was about as long and wide as a shipping container, and each step I took crackled with the sound of crushed plastic. I pulled out muddy, sodden clumps of Styrofoam, slimy bottles, and slippery buoys. One teenage volunteer had followed Mackenzie and me into the brush and disappeared into the undergrowth. He returned with five garbage bags crammed full. I tried to imprint the image on my brain but pulled out my phone and snapped a picture of the forest floor, just in case my memory failed.

Mackenzie told me this wasn't even close to the worst she had seen on the trip. A few days before I arrived, she cleaned a beach littered with tiny plastic pieces. Much of the visible plastic pollution that

Ocean Legacy collects in MuQwin washes up in whole chunks that are at least recognizable as a milk jug or a Gatorade bottle. But this beach was covered in pieces no bigger than a fingernail. Ocean Legacy follows the National Oceanic and Atmospheric Administration's cleanup guidelines and collects anything bigger than the size of a bottle cap, so Mackenzie crouched down and started sieving plastic from sand. Three hours later, she was still in the exact same spot. Her garbage bag was only half full of thousands of plastic chips. The sand beneath her feet was still thick with thousands more multicolored fragments. On every expedition, the team came across at least one staggering discovery like that. A few days after I left, they found another: a stretch of beach covered with thousands of individual broken-apart pellets of foam buried six feet, as deep as a grave, through the sand.

As I hiked through the plastic-covered swamp, I began to experience the same phenomenon that Chloé had seen out in the Great Pacific Garbage Patch. A flash of white would catch my eye. I would stop to pull a crumbling piece of foam from the forest floor, only to realize that a sapling had taken root in the foam pellets. One volunteer told me about a thicket he had found full of mushrooms sprouting through plastic mulch. Another emerged from the forest holding aloft a piece of foam with claw marks deep in its sides—a bear cub's scratching post, perhaps. Chloé told me about a wolves' nest she found built from foam, a first-rate insulator. On land and at sea, the animals were working with the plastic. But those revelations were the happy exceptions. So often the interactions between plastic and animals ended badly—for the animals. After watching a now infamous video of a sea turtle with a plastic straw jammed up its nose, Chloé was never able to look at straws the same way again.

Straws are relatively minor compared to the damage done by so-called ghost gear. It's believed that around ten percent of plastic in the ocean is abandoned or lost fishing equipment that, long after any human fisher has finished with it, continues to collect bounty. All types of animals, from birds to sea turtles and invertebrates big and small, are snared in drifting nets or caught and killed in abandoned traps. When that animal dies, it sets off a truly depressing cycle of events, as its carcass lures in more animals looking for an easy bite. The cycle repeats.

Ghost gear might not be a modern development. Accounts of aquatic monsters pepper captains' logs dating back to the 17th century. In 1875, Captain George Devar and his crew spotted a sea serpent devouring a whale off the coast of Brazil, or so they thought. A far more likely explanation is that the whale had lost a battle with a discarded fishing net. Once upon a time, such stories might have stirred up imaginings about the age-old mystery of the sea. Today, the industrial levels of plastic pollution has accelerated the number of encounters between plastic and animals, making them prosaic and deathly. In 2014, the World Society for the Protection of Animals estimated, conservatively, that ghost gear killed 136,000 seals, sea lions, and whales every year. Researchers theorize that modern plastic fishing equipment can drift for up to six hundred years in the ocean before breaking down.

An hour into clearing the swamp, Mackenzie and I decided to limit our sweep to the accessible areas bordered by reeds and brushes. Otherwise we would be bushwhacking for days and probably do more harm than good. It occurred to me that all the forest just beyond the shores of MuQwin that I had cleaned were probably just as packed with plastic, but the brush was too dense to penetrate. I hauled out my final garbage bag, full of plastic, knowing that so much more plastic was spread out across the forty thousand hectares of MuQwin Peninsula, and that there were many tons of plastic along coastlines and in waterways around the world. I felt deeply tired.

After climbing out of the swamp, I showed Chloé the picture I'd snapped. "Would anyone believe this much plastic existed here?" I asked her. Often, she told me, the answer is no. On Facebook, when she posts about her cleanups, she frequently encounters disbelief.

Videos she shoots of foam pellets spread through the sand, or a time-lapse of volunteers emptying out a crevasse packed with plastic, are met with incredulity. One older man who had grown up in MuQwin Peninsula commented on her videos that he remembered a more pristine place from his childhood. He probably had experienced an earlier, cleaner time in MuQwin's history, but it wasn't the reality anymore. She invited him to come clean with her. He never wrote back.

Before plastic pollution was considered a serious environmental threat, whistle-blowers often faced disbelief and even hostility. Scientists initially doubted the British sailor Charles Moore, who reported his discovery of the Great Pacific Garbage Patch in the mid-1990s. Chris Jordan, a Seattle photographer, became famous for his images of dissected albatrosses, their stomachs full of bottle caps and plastic chunks. He's been accused of staging his work, and now he posts start-to-finish videos of the dissections.

It took some imagination to square my clean city streets with the destruction I saw in MuQwin, because we intuitively think of rural or remote landscapes as cleaner than urban ones. In fact, the opposite is often true. Most of us are used to street cleaning, garbage pickup, laws that ban littering, and officers to enforce those laws. We live in cities and towns and care about having clean parks, beaches, and streets. This managed sanitation cuts us off from what's happening out at sea, where no one is present to stop or even witness the pollution.

In the age of climate change denial, many scientists and activists often confront a willful disbelief about the threatened state of the ocean today. Environmental damage is often too slow, too scattered, too distant to instill any sense of urgency in the skeptical or disengaged. Even when those impacts are condensed into hard statistics and diagrams or communicated through rigorously documented studies and video evidence, people can simply choose to believe their own particular truth.

When I climbed out of the swamp, I tried to explain to Chloé the sort of general sense of sadness I felt. I had never felt this way looking at a landscape. Now I wished I hadn't seen the swamp, that I could go back to the cleaner, more manageable beach of yesterday and the easy ignorance that plastic pollution wasn't *that* bad. Chloé understood how I was feeling. A few years earlier, she'd taken a course by the eco-philosopher Joanna Macy, and she'd seen military veterans—"people who had never cried a day in their life"—openly weep as they'd spoken of the destruction they had experienced or caused. "In our culture there's no space for that," she said. "When you see a tree being cut down, and you feel something for that, most of the time it's mocked in our Western culture."

As Chloé talked about eco-philosophers and the energy of environmental PTSD, I heard skepticism in my own head. *Some hippie stuff right here*, I thought, just as the mainstream dismissed crying-Indian ads, tree-hugging hippies, and bleeding-heart liberals. This emotional, unscientific approach to environmentalism tended to turn people off, myself included. Even though the pollution I saw stirred a deep feeling of sorrow in me, I wasn't ready yet to center myself as a victim of environmental trauma. What good were emotions when systemic forces were destroying the ocean?

Months after I returned home, I found that my sadness about MuQwin Peninsula wasn't going away. When I looked around, I saw plastic following me everywhere I went, and with everything I did. I tried to police just how much plastic slipped through my fingertips. At the deli, the butcher wrapped my meat in plastic even when I asked for paper. On the train, the cafe refused to fill the reusable mug I brought because it violated their safety rules. At the grocery store, I paid almost triple what I would for canola oil when I refilled my own container. All this plastic felt exhausting and expensive to control, as well as completely beyond the average person's time and budget. I felt annoyed, then angry, and finally just powerless to stop the catastrophe. On top of that I felt a profound guilt that people in poor, underdeveloped countries would bear the worst effects of plastic pollution.

They would also bear the worst effects of climate change. In the Pacific Ocean, communities on small island states, like the Marshall Islands, sit less than a couple of meters (five feet) above sea level. When big waves hit, buildings are wrecked, crops die, and the ocean seeps into the water table, making the water too salty to drink. The current generation will likely be the last to live on these far-flung atolls. On the opposite end of the world, year-round Arctic ice has shrunk drastically over the last decade. In northern Inuit communities, that ice is a hunting ground, a highway, a bridge between far-flung villages and territories. The Inuit writer Sheila Watt-Cloutier has watched closely as a mental-health crisis has gripped her people, and she links it to climate trauma. Losing ice means losing an identity, a livelihood, and a way of life, all at once.

I started to keep a growing collection of stories about environmental trauma, more commonly known as climate grief. I read a conceptual paper on societal collapse so unsettling that it spawned an international network of climate grief support groups. I read about biologists who broke down in tears after seeing the impact of the latest coral-bleaching event in the Great Barrier Reef. I read about natural disasters, like Hurricanes Sandy and Katrina, and how the communities these hurricanes affected went through a spike of mental-health crises. As we enter a more unstable future, where hundred-year floods and billion-dollar hurricane damages become the norm, we're bound to see more cases.

In this darkest of places, I saw a way to save the ocean. Just as we ascribe economic value to a tourist beach or a fishing industry, we can give emotional value to a healthy ocean. When climate deniers dismiss environmental destruction, we can use collective grief to prove that it has a real impact. This is a new frontier in saving the ocean.

Mackenzie, the chef who showed me the plastic jungle, told me about one keepsake she found: a tiny bottle with a scroll inside.

She unrolled it and read, "In wilderness is the preservation of the world"—one of Henry David Thoreau's most frequently quoted lines. She kept the scroll with her as we picked plastic from the sand.

~

"We look like we're driving off the edge of the Earth," said James. Through the windshield of the *Imperial*'s pilothouse, we looked out at a wall of whiteness. The trawler was leaving the quiet, wind-sheltered channel where we had camped and heading out to open water. That morning I had peeled back the flaps of my tent opening to see an all-day downpour beginning—typical weather for the Pacific Northwest. As we left the channel, the weather went from moody to outright angry. A thick fog surrounded the ship on all sides. James wiped water droplets and steam from the windshield. Perhaps five meters of visibly choppy ocean lay ahead as the boat pitched violently from side to side.

My team was heading back to Kyuquot. We would return to the mainland, and a new group of recruits would take our place. The fierce weather seemed a perfect way to end the week. Each day, I had hauled something surprising from the sand and discovered yet another angle about plastic's relationship with the sea. In the end, I wasn't disappointed by the amount of plastic I'd found. What we collected was only a fraction of a much bigger dump unseen out at sea.

The bow of the *Imperial* bucked high into the air, and the waves shuddered through the pilothouse. Neither James nor Chloé reacted to the wild swings the ship made every few seconds. James worked the helm; Chloé scrawled the day's events in a binder. I had never sailed in worse weather, and I was struggling to keep my nerves in check. Rain leaked down through the peeling wallpaper of the *Imperial*'s wheelhouse and splashed onto my notebook.

As the *Imperial* nosed its way farther out to sea, I thought about a question I had refrained from asking Chloé all week. In the coming month, she and the rest of the Ocean Legacy team would transport over twenty-five metric tons of plastic debris back to the mainland. Many cleanup operations landfill most of what they find, but her team was going to recycle and repurpose the entire haul. The process would take months of planning and coordination, not to mention the six weeks of collecting and the labor of more than two dozen people to untangle plastic from the shoreline. Every day, that same amount of plastic and more was dumped back into the ocean—around twenty dump trucks' worth. How depressing to realize that your work is being undone before you finish it. Did she ever feel frustrated that all her efforts might not save the ocean?

"I don't get frustrated by it," she said, "but the thought is daunting. It becomes more daunting when you think of the scope of the planet. There're probably millions of beaches around the world going through the same thing as MuQwin Peninsula." With a little money and a team of die-hard plastic collectors, Chloé had found a way to press forward.

Sometimes when Chloé was digging through the sand of MuQwin Peninsula, removing plastic foam pellet by tiny pellet, she thought about what caused us to release plastic so unconsciously into the natural world. Partly, it's convenience. We can continue to dump plastic for centuries before we'll run out of space—that's how good the ocean is at hiding our secrets. Compared to ocean acidification and warming, plastic pollution is one of the more visible problems facing the ocean today because almost everyone, every day, throws away some piece of plastic. This makes it hard to eliminate, but also more visible and accessible. We have a greater chance of confronting this problem because the plastic is in everyone's hands.

Dragging a garbage bag full of plastic on the sand became my own personal penance for all the plastic I had thrown away in my lifetime. That week ensured I would never again throw away a plastic Q-tip so recklessly. Fixing the larger issues of protecting the high seas feels more remote. It means undoing how we think, or don't think, about the ocean. It means cooperative regulation and reprimanding bad actors. It means tackling one of the hardest challenges that has ever faced humanity. The ocean is begging for our imagination and cooperation.

On that rocky boat ride out of MuQwin Peninsula, I thought about a moment when this place gave me a chance to look beneath the surface—and not just the ocean's plastic pollution. A few days earlier, when the team was riding back to camp, everyone exhausted from a long day in the sun, a volunteer spotted a killer whale. "Orca!" he shouted.

Those two syllables had an electrifying effect on the crew. People who had passed out on deck suddenly sprang to life and ran to the *Imperial*'s rails. A couple dozen meters from the boat, four black fins broke the surface. Four blowholes sprayed salty mist into the sunlight, leaving the gauzy arc of a rainbow. For a few moments, everyone stood in awe. This was the closest I'd ever come to a wild whale, and here they were, swimming past a load of plastic. For those moments, it was unwaveringly clear why Chloé and her team were here, removing plastic from the ocean, one bottle at a time. Then the whales took a breath and disappeared down below.

SIX

CRUISING THE NORTH ATLANTIC

"Some weeks before I underwent my own Luxury Cruise, a sixteen-year-old male did a Brody off the upper deck of a Megaship—I think a Carnival or a Crystal ship—a suicide. The news version was that it had been an unhappy adolescent thing, a shipboard romance gone bad etc. I think part of it was something there's no way a real news story could cover."

—David Foster Wallace

It was Friday night in the crew bar of the *Queen Mary 2*, and the cruise ship's waiters and chefs, busboys and cleaning ladies, lifeguards and luggage stewards were dancing and drinking. Early tomorrow morning, they would rise from their bunks, put on their uniforms, pin their name tags to their chests, lock their tiny cabins, thread through the long, windowless corridors below deck, and take their positions for the long day ahead. Favio Oñate Órdenes, a 26-year-old chef, weaved through the crowd. He was a regular fixture at the crew bar, volunteering many nights as a DJ, drinking and hanging around until late. At 1:00 A.M., the crew bar shuttered for the night, and most of the staff dispersed back to their cabins. But Favio and two other crew members didn't feel like turning in just yet. Balancing their drinks, they went outside to the crew deck, to smoke and talk a little while longer, as the Atlantic Ocean passed, big and dark before them.

Favio was twenty-three when he left his life in Chile for a chance to see the world. In the three and a half years since, he worked, slept, socialized, and came of age on the ship as it crisscrossed the globe. He visited dozens of port cities, racking up more nautical miles than most of us will ever manage in a lifetime. Working on a cruise ship had seemed like a life-changing opportunity when he'd worked in a casino kitchen on the outskirts of Chile's smoggy capital, Santiago. He had paid off his student debts—slowly—and sent his family money. He wanted to help out even more, particularly because his younger brother had diabetes and required expensive insulin injections. Working on a cruise ship promised double the average pay a chef could earn in Chile, plus the opportunity to travel and get a well-established and internationally recognized company on his resumé. His English was strong from a year apprenticing at a restaurant in Vermont after he finished his degree. Less than two weeks after his interviews with Cunard, he had a six-month, seven-days-a-week position on the *Queen Mary 2*, a contract that was renewed multiple times over the coming years. He packed his bags and went to sea, leaving behind his family and friends.

Favio was excited to earn more money, meet new people, and travel farther than he ever had on land. But when he cut a deal with the cruise ship, the exchange he made was much like the contracts sailors have always made with the sea. He traded his time and skills, and he joined a largely invisible world of workers on the ocean.

The *Queen Mary 2* must have impressed him when he arrived at the dock on his first day. In nautical circles, she is a marvel. As long as three football fields, as tall as the Statue of Liberty, she is one of the most lavish, most elegant, and most spacious passenger ships cruising the seas today. Launched in 2004 and marketed as "The Last Great Ocean Liner," she was designed to evoke the golden age of ocean travel a century earlier, and named to honor RMS *Queen Mary*, which for three decades provided elegant, midcentury passenger service across the Atlantic (RMS *Queen Mary* is now permanently moored and a tourist attraction in Long Beach, California.) *Queen Mary 2*'s hull is painted not the garish white of modern superships but the nautical blue and red of steamships past. The ship's officers dress in white, their shoulders trimmed with gold admiralty epaulets. Cunard names all its modern ships for British royalty, perhaps because it appeals to the older anglophile cruisers who appreciate Cunard's storied maritime history. The interiors are decorated in nautical British style that evokes the grandeur of the *Titanic*, the most infamous and elegant of all ocean liners. Amenities include a library, a bookshop, and a ballroom, where white-gloved waiters serve afternoon tea. The dress code is country-club casual, and passengers dress up for dinner. Along the decks, people play shuffleboard or gaze out over ocean horizons.

Each year, the *Queen Mary 2* embarks on a four-month world voyage that starts in Europe. The ship threads its way through the Mediterranean and the Suez Canal, into the Bay of Bengal, around Australia, across the Indian Ocean, and back again to Europe. Favio posted many photographs on his Facebook page: Favio holding up a distant Egyptian pyramid in each palm; Favio hugging a camel in Dubai. But the ship is most famous for its seven-day transatlantic crossings from Southampton, where Cunard is headquartered, to New York City. From spring until fall, the *Queen Mary 2* crosses the Atlantic over a dozen times, stopping in ports in Iceland and Canada along the way. Favio had watched Sting play on board, and Blink

182's Travis Barker, too, who jammed with the crew bands. Favio loved New York City. He told his mother that he teared up at the sight of Central Park blanketed in snow, just like he remembered it in *Home Alone*, one of his favorite childhood movies. In Hell's Kitchen, Favio posed for pictures with his arms flung wide, imitating the awe-struck sailors in the Gene Kelly musical *New York, New York*. "My place of battles, sorrows and joy, my apprenticeship, my future, my big step and the opportunity I always wanted," he wrote on Facebook next to a picture of the massive ship.

Favio spent his first contract tucked far behind the scenes. He worked scattered early-morning and late-night shifts in the galley, cooking endless pots of rice for the crew. Although the work didn't challenge him, he kept up a chatty disposition in the kitchen. Most of the other kitchen workers were Filipino, and he learned a few Tagalog phrases and inserted Chilean dialect to make his Filipino coworkers laugh. During his second contract, Favio left behind the steaming pots of crew rice to become *commis*, an apprentice chef who circulates through the kitchen learning different roles in different restaurants. He was excited to cook desserts for the passengers. He learned how to carve giant ice sculptures. His work hours improved, too. Instead of the numbing routine of moving back and forth between day shifts and night shifts of varying lengths, he worked longer hours during the day, similar to what he'd worked in kitchens in Chile. With his evenings free, he planned to go out partying more and meet other crew on the ship.

When Favio joined the crew of the *Queen Mary 2*, he met Santiago Gutiérrez, who was also a Chilean chef working his first contract with Cunard. Santiago started out working in the King's Court, the cheapest buffet restaurant on board. All day he stood behind steaming hot plates and chatted with the ship's rotating passengers, who assumed he was French, with his light skin, black hair, and accented English. He dreamed of working in the flames and frenzy of the "real" kitchen down below.

Not long after he started, a chef called to him across the galley, "Hey, one of your *paisans* is coming." This meant a new hire from Chile, and that was notable. No more than ten Chileans were on the whole crew, which was more than a hundred times that number; for most of their time together on board, Santiago and Favio were part of a small contingent of Chileans, three or four at most, in a galley of predominantly Filipino chefs. They became a sort of family at sea, even though they looked and acted nothing alike. Favio was short, muscular, and friendly; Santiago was tall, thin, and outspoken. But they were like brothers in many ways. They did everything together: working out, partying, sightseeing around the world. If anyone or anything ran afoul of Favio, Santiago heard about it and vice versa.

When they met for the first time, they played the name-game, searching for similarities that bound them beyond nationality. Both had studied culinary arts at INACAP, a technical institute with locations across Chile. Both came from close-knit families, and they grew up in cities fifty kilometers apart. After they earned their diplomas, both worked a year abroad in North America—Santiago in Canada and Favio in the United States. Both returned to Chile to work, and both ended up applying to Cunard.

"We just grabbed the friendship," remembered Santiago. It was natural to cling to someone who felt like home. They moved into a cabin together and synced their contracts so that they would always share a space with a friend rather than a randomly assigned roommate. "Nobody can understand the strong connection you can have with someone you're living with in this kind of environment for three years."

Cruise-ship workers often report feeling lonely at the start of their new life on board. They're far from home. They might have to adjust to a new or unfamiliar language. In the crew corridors, the industrial hallways look like miniature airport runways. Yellow and black arrows painted on the floor direct foot traffic. The walls are made of plates of corrugated metal bolted together and painted white. Shin-high yellow guardrails prevent industrial equipment from scraping the walls as it passes through. There are no windows, so there is no natural light. Blinding fluorescent strip lights guide the way. New crew members have to learn how to navigate the massive ship so they can get to their shifts on time. They desperately don't want to screw up the many spoken and unspoken rules: always smile at the passengers, but don't sleep with them (that's a fireable offense). Then there are all the hidden hierarchies and dynamics of a workplace to navigate. The long days, tight quarters, and culture clashes on a cruise ship only exacerbate these tensions.

During one contract on the *Queen Mary 2*, a woman Favio had become involved with on a vacation back in Chile broke things off with him. The distance and time away were simply too much to handle. "Favio was so sad," Santiago remembered. He cried through his shifts. At the time, Santiago and Favio were working at stations near each other in the Britannia restaurant on board. A few of their coworkers approached Santiago and told him to keep his friend in line. It was fine to be sad, but no one wanted to see a coworker weeping over a pot all day. Many of the crew had relationship or martial problems, suffered from homesickness, or had children they never saw.

Working in the belly of the ship, the kitchen staff rarely saw the outside world. The *Queen Mary 2* sailed over the wide-open ocean and past exotic ports, but inside, the fluorescent-light existence made it hard to keep track of time and space. On her transatlantic crossings, the ship sheds or gains an hour every day, depending on the route. The crew begins to adjust to a perpetual feeling of floating between time and place.

Santiago worked ten-hour days, seven days a week, and received the occasional overtime pay on *Queen Mary 2*. There are reports, however, that overtime goes unpaid on other cruise ships. In the magazine *California Sunday*, ProPublica reporter Lizzie Presser wrote of a Filipino waiter on a Carnival ship who worked as much as fourteen hours some days, but the time-sheet software only allowed him to log the official ten. Ten other Carnival employees Presser interviewed had the same experience.

In the close confines of the vast ship, coworkers keep a watchful eye on one another. Rumors spread through the cabins and decks the way gossip rips through a small town. "As soon as you make a mistake, everyone knows. As soon as you hook up with someone, everyone knows," Santiago said of the *Queen Mary 2* fishbowl. "You cannot hide yourself."

"You need to change your act; you always look angry," Santiago remembered Favio warning him. But Santiago cautioned Favio to stay on his guard around his coworkers. "You can't be nice with everybody," Santiago said. "There are people you joke with, but as soon as you turn around, they're gonna stab you in the back."

Over the shipwide network of CCTV cameras, managers had the ability to wield unprecedented surveillance over their employees' lives if they wanted. On the decks, in the gym, in every corridor, and nearly everywhere workers went, except for their shared cabins, the washrooms, and the galleys, according to Santiago, their bosses could see exactly what they did—on and off the clock. Crew cleaned their cabins every two weeks for inspection by upper management.

At night, in the tiny cabin Santiago and Favio shared, they had some privacy. As they drifted off to sleep in their bunks, they talked about the future, imagining what their lives would look like off the ship. Many chefs dream of starting a restaurant one day—it's natural, given the militant environment of a cruise-ship kitchen—and they shared that dream, too. More often their talk turned to life outside work, to escape. One day they wanted to visit Ibiza together. What would it feel like to be friends in a world where every minute of their day wasn't rigorously observed and counted?

Three years into his time at sea, Santiago's father suffered a heart attack in Chile. As he got off the phone with his sister, all the fatigue and stress of the past three years caught up to him in an instant.

When he went back to the kitchen and his boss swore at him, he finally snapped. "Only my dad is allowed to use that language with me!" he shouted back. The galley went silent, every worker staring at him in disbelief. This was in the early spring of 2015, when Favio was back in Chile on a vacation between contracts. Santiago sent him a message on Facebook to tell him he was leaving the ship. Favio was getting ready to return, so their paths wouldn't cross again. Maybe they would meet in Ibiza someday.

Now, in the early hours of Saturday, August 15, 2015, Favio was out on the crew deck, drunk, talking loudly, and even shouting at times. Moments before, he had disappeared into a washroom with a male coworker. He emerged five minutes later, transformed. He was visibly shaken, but the other crew member who went into the washroom behaved exactly the same as before.

The *Queen Mary 2* was halfway through its popular transatlantic summer cruise. Over 750 kilometers away lay the coast of Newfoundland, Canada. Suspended between two continents, civilization far off in the distance on either side, the 345-meter ship was nothing but a speck of electric light in a sea of darkness. A crew member would later tell reporters that Favio announced he was going to jump off the ship that night. If Favio did make such an announcement, it apparently didn't make a big impact on the people he was with. The crew bar is often the location of many outrageous declarations over the years: people swore they were going to quit, beat up their bosses, or tell off their roommates as soon as they got the chance. The crew bar serves as the pressure relief valve on every ship. It's the one place where overworked and underpaid staff can blow off steam. At the time, Favio probably looked like he was destined for a bad hangover and an equally bad day in the galley tomorrow. At most, his antics might be a good story to tell the following night at the crew bar.

Favio's coworkers tried to take him back to his cabin and put him to bed. But he resisted, and eventually the crew members left him alone, sitting on the floor of an elevator by himself. At some point,

Favio got to his feet. He took off his shoes and a neck chain, one that he never removed, and left them in a stairwell as he roamed the empty corridors of the *Queen Mary 2* by himself. Outside the darkened windows, the Atlantic Ocean slid past. It was late now, nearing 3:00 A.M., and in just five hours Favio would be back on the line again, cooking breakfast for the passengers. Out of uniform—a serious violation of ship policy—he went out onto passenger decks. He leaned against the railings, standing where hundreds of passengers had stood before him, and looked out at the ocean.

He wouldn't be seen again.

On the morning of Saturday, August 15, the passengers on the *Queen Mary 2* woke to a strange announcement: the deputy captain asked *commis* chef Favio Oñate Órdenes to report to the executive chef. His shift had started at 8:00 A.M., but over an hour had passed and still he'd failed to appear. Three more public announcements were made as staff searched his cabin and questioned his roommate. Security scanned the surveillance videos recorded in every public nook and cranny on the ship. At 10:13 A.M. they discovered footage, time-stamped at 2:47 A.M., of a man falling overboard from deck 7. Captain Kevin Oprey altered the ship's course, motoring back to where she had sailed the night before. Mayday calls went out to ships in the area, and a container ship on its way to Montreal turned back to take part in the search. According to tweets from passengers on board, the ship announced that the missing crew member had fallen overboard.

That morning, a Miami lawyer received an anonymous phone call. The caller—a crew member who did not want to be named—told maritime lawyer Jim Walker that the *Queen Mary 2* was searching for a young Chilean chef lost at sea. Such phone calls are not unusual for him. Walker runs the blog Cruise Line News (tagline: "Everything the cruise lines don't want you to know") and invites readers to send him tips. Anytime something untoward happens on a cruise ship, he is one of the first people on land to hear about it.

Shortly after the call, Walker posted the story to his blog. He headlined it CUNARD CREW MEMBER LOST AT SEA. From there, the online community of cruise-ship workers picked up the story and shared it across social media.

That Saturday morning, around nine thousand kilometers to the south of the ship, Santiago Gutiérrez turned on his cellphone and received a flurry of text messages asking if he was okay. A friend of his had just read about a Chilean chef who had disappeared from the *Queen Mary 2*. The friend didn't know that Santiago had quit the ship five months earlier and was now living at his parents' house. She wanted to confirm that Santiago was okay. He searched online and found Jim Walker's story about a Chilean chef lost at sea. The blog post didn't mention a name, but Santiago knew who had gone missing.

Thousands of miles away, Santiago received updates from friends and former coworkers still on the ship, and he felt responsible. He couldn't do anything to help a friend who had been his support system for years, he realized, but he could help Favio's family. Their English was limited, and they didn't have the same contacts on the *Queen Mary 2* that Santiago had. They might not know yet that he was missing. He asked his friends on board whether anyone from the ship had notified Favio's family. No one had. Santiago contacted Favio's sister, Carolina Oñate Órdenes, through Facebook and asked for her phone number. He shared with her a link to the *Queen Mary 2* Facebook page, where, at 1:12 P.M. Chilean time, the ship had reposted a friend's farewell message to Favio, alongside a picture of the young chef standing with three carved Halloween pumpkins.

Favio's sister had the unimaginably difficult job of telling her mother that her son was missing. Carolina Órdenes Tobar was away on a vacation when her daughter called, and she refused to believe the news at first. She spent the rest of the afternoon vacillating between fear and disbelief. The post must have been a sick joke, or maybe it didn't concern her. Surely a big company like Cunard wouldn't post a public message about an overboard without first informing the family.

A month earlier, Carolina had spoken with Favio over Skype. (Phone cards are expensive, so mother and son usually spoke through online apps.) She remembered nothing extraordinary about the maternal check-in. He'd sounded like his usual happy self. He'd been excited about winning a bet on a soccer game between Chile and Argentina. He'd asked about his younger brother's grades. His sister, too, had spoken with him on Skype just two days before he disappeared. She remembered he'd looked happy. He'd told her that with his new promotion and improved work hours, this was his best trip so far.

Throughout the afternoon, the family tried to contact Favio—no response. They called and left messages with Cunard and messaged the company through social media—nothing.

Around noon, the *Queen Mary 2* increased speed, returning to the stretch of water where the ship officers calculated Favio might have drifted. The ship started the search at 5:00 P.M.—over fourteen hours since Favio had gone overboard at 2:47 A.M.—and would continue searching the water until dark. The water temperature was around 12°C. Someone might survive six hours at that temperature, but that didn't take into account the nearly ten stories Favio fell from deck 7 down onto hard ocean. The *Queen Mary 2* and the passing container ship started running search tracks parallel to each other as twilight approached. Passengers posted to their Twitter accounts photos of people crowding the railings, looking for Favio as a dense fog enveloped the ship. An hour into the search, something was spotted in the water, and the ship turned to investigate. It was a false alarm. The ships went back to tracking the same search pattern. The daylight began to dwindle, the fog grew thicker, and the chance of finding him faded.

The chances of recovering a person lost at sea are nearly impossible. When Malaysia Airlines flight MH370 crashed into the Indian Ocean in 2014, it seemed astounding that the best minds, resources, and technology on land couldn't find a plane lost at sea. But as of this writing, only a few pieces of MH370 wreckage have drifted onshore, and the plane's exact resting spot is still unknown. That was a commercial jetliner almost forty times the length of a person. Even a hundred meters of choppy ocean can make it difficult to spot a human head.

That evening, a representative from Cunard reached Favio's sister at her mother's home. The captain had called off the search for Favio.

❧

Humans cross the ocean for many reasons: food, travel, exploration, exploitation, conquest, colonization. One thing holds true about sea crossings in the past: they were not fun. "Those who would go to sea for pleasure would go to hell for pastime," one English seaman's proverb put it. We don't know when exactly the first people took to water, or why they did it, but the ocean was simply too dangerous, too difficult, and too mysterious to cross purely for pleasure or the sake of curiosity. The early Polynesians' advanced seafaring knowledge allowed them to navigate by the stars and swells and delineate the loom of land thirty miles in the distance. Their purpose for crossing vast swaths of the Pacific is still unknown. A plan to cross the ocean raises an obvious question: how do we return? Europeans who set sail in the 14th and 15th centuries understood that the circular trade winds of the Atlantic would carry them home again, just as the early voyagers near the Solomon Islands worked nearby winds to similar effect. Diverse and sometimes overlapping cultures lay claim to advances in navigation and maritime technology, but it was the cruise industry that played a leading role in creating a comfortable, even elitist, experience of travel for travel's sake. From the late Victorian era up until the 1950s, ocean liners captured the public imagination with their opulence and speed. The French Line had its art deco–designed *Ile De France*; Cunard had *Mauretania*, which set repeat records for swiftest transatlantic crossing. In the gilded age of ocean liners, it's hard not to see countries, and their cruise ships, flirting with the disastrous nationalism that would dominate the early 20th century. Only recently, with the rise of booze cruises, party boats, and harbor tours, have we begun to associate the ocean with conga lines and bottomless margaritas.

Cunard has run ships across the Atlantic since 1839. After the *Titanic* sank in April 1912, Cunard's RMS *Carpathia* was first on the scene and pulled survivors from the *Titanic*'s lifeboats. Three years later, a German U-boat torpedoed one of Cunard's biggest and grandest ocean liner, the *Lusitania*, killing nearly 1,200 and eventually bringing the United States into World War I. Winston Churchill credited Cunard's transatlantic "grayhounds" with shortening World War II by at least a year by transporting Allied infantry across the ocean. By the late 1960s, however, fast and affordable air travel rendered Cunard's ocean liners increasingly obsolete. Cunard had to offer something more than an ocean crossing.

In the 1970s, the industry introduced the cruising we know today: gleaming white mammoth ships with water parks and planetariums on board. Cruising was about the journey, not the destination, and the popular American TV series *The Love Boat* both reflected and reinforced cruising's appeal. "Wall Street got into [cruising] in the mid-1980s," observes Ross Klein, a sociologist who is an expert on the cruise-ship industry. "That's when you see a major growth, a burgeoning in terms of larger ships."

Throughout the 1990s, Cunard's ships slowly became a part of the Carnival empire, one of the three corporations, known as "The Big Three," that control cruising today. Together, Carnival Corporation, Royal Caribbean Cruises, and Norwegian Cruise Lines Holdings operate 155 ships across eighteen subsidiary brands. In 2015, they made a combined profit of $2.8 billion and dominated 82 percent of the market.

The size of the ships, the number of passengers, and the features and options have all exploded in recent years. The ships are so large that they can carry the population of a small town, as many as six thousand passengers, but with better amenities and attractions. Ships offer outdoor water parks with massive LED screens suspended above them—a weird mash-up of Times Square with a pool and an ocean. Go-kart tracks, 5D cinemas, planetariums, sporting arenas, rock-climbing walls, waterslides, Guy Fieri burger joints, and a range of entertainers are the norm.

Cruise ships fill a certain niche in the tourism world. They appeal to families and newlyweds; to people who hate flying or just find it uncomfortable, like the elderly and the overweight; and, above all, to people who want a luxury vacation without the price tag. (There's an unkind saying that sums up the type of passenger cruise ships serve: newlyweds, overfeds, and nearly deads.) In the last three decades, the Big Three have duked it out for customers in an increasingly cutthroat world, battling for market share with the lowest price tag. A cruise-ship passenger sailing from Vancouver to Alaska could pay as little as $33 per day in 2019. That includes accommodation, food, entertainment, day care, and some on-board and port-of-call activities. The price of a basic hotel costs more than that per night without including any extras.

The *Queen Mary 2*, of course, is different. It offers traditional ocean-liner service in this new mass market of floating theme parks. Repeat travellers call themselves "Cunardists," and the trip is always a crossing, never a cruise. The *Queen Mary 2* is quiet and spacious compared with other cruise ships; there is a library, a bookshop, and a planetarium. The captain gives only one announcement a day. The cheapest fares on board average around $175 per day and run over $600 a night for a lavish suite. Among the monster ships of today, the *Queen Mary 2* has one of the more generous staff-to-passenger ratios in the industry, with 1,292 employees serving 2,691 passengers, according to Cunard's website. After all, someone has to prepare the food, lifeguard the pools, staff the shops, administer spa treatments, and watch over the kids. With the growth of cruise ships has come a corresponding growth in labor—and labor can be expensive.

When the British Admiralty granted Cunard the first transatlantic mail-carrying contract in 1839, the agreement stipulated that three quarters of the crew be British subjects. Ocean liners were once nationalistic entities: Greeks ran the Greek Line, Italians the Italian Line, and so on. Today, every one of the Big Three parent companies is based in Miami, Florida. The majority of cruise-ship workers are Filipino, according to researcher William C. Terry, while other low-ranking jobs are filled by workers from India or Indonesia, according to a paper published by Bin Wu at Cardiff University's Seafarers International Research Center.

In 2016, an investigative reporter from Univision Noticias, an American-Spanish language broadcast network, described going below deck on a Carnival ship to see where staff live and work. "Indians dominated the kitchen, Indonesians were in the refrigeration and laundry areas, Filipinos in the hotel rooms, Caribbean and Central Americans in maintenance, Italians in the control rooms, and at the casino, Eastern Europeans cut the decks of cards . . ." the reporter wrote in a sweeping investigation entitled "Sweatshop on the High Seas." He also reported that American staff, as well as anyone from a developed nation, worked in a higher public-facing position, such as a senior officer or daycare supervisor, and had shorter hours for better pay. Newspapers like *The Guardian* and academics like Bin Wu and Christine Chin have noted similar practices. ProPublica reporter Lizzie Presser traced the hiring practices of cruise ships to the colonial ties between countries in *California Sunday*. "Holland America, a Carnival subsidiary that was once Dutch, runs training schools in its former colony of Indonesia, and 48 per cent of its crew members are Indonesian. P&O Cruises, another Carnival subsidiary based in Britain, has recruited in India since the 1850s," she wrote.

Earnings differ sharply between cruise-ship employees. In a 2019 story, *Business Insider* interviewed 35 current and former workers and found that pay can range from $500 to $10,000 a month. The average salary, however, lies somewhere between $1,000 to $1, 500 monthly, while behind-the-scenes labour tends to earn less, according to the reporting from Univision Noticias.

South American workers like Favio and Santiago worked six-month contracts with no days off. At certain ports, the crew occasionally get a few free hours to wander the city, but mostly they work around the clock—Sunday like Monday like Tuesday like Wednesday like Thursday like Friday like Saturday like Sunday. Santiago remembered that the Filipinos worked nine-month contracts with no days off and were paid less, even when they filled the same positions as South American workers. Santiago recalls working next to a Filipino chef who had worked on the ship for five years. He'd worked his way up to earn as much as Santiago was paid in his first contract. According to Santiago, he and Favio earned around $2,000 each month, while most of their Filipino colleagues earned about half that much. "That's why they don't like us," Santiago said simply.

According to Santiago, it wasn't only unequal wages that grouped employees along ethnic lines. When one Chilean created problems, or couldn't keep up, Santiago would hear about it, and that made him feel responsible for the other Chileans on the ship. He worried when they complained too loudly about the long hours, or the shouting, or the chefs stealing prepped food from each other. His perspective, common among crew, was: if you don't like it, fly home. Invariably, another staff member would ask him to get his *paisano* in line and explain the White Star service and standards that the *Queen Mary 2* required.

The employment standards that govern Cunard's headquarters in the United Kingdom don't apply to its staff at sea. The reason why lies on the *Queen Mary 2*'s stern: Hamilton, Bermuda's capital, is painted across the back of all three Cunard ships. According to the United Nations Convention on the Law of the Sea, every vessel must fly a flag that shows where in the world she is registered. When ships sail out of a country's territorial waters, they enter international waters.

The flag they fly signals that, legally, the ship's crew is working in that country and is subject to its laws. Even if an accident occurs when a ship is docked in port, the crew is not necessarily protected by that country's laws. A watershed case occurred in 2003 when a boiler exploded on a Norwegian Cruise Lines ship, killing eight crew members and injuring another ten, while the ship was docked in the Port of Miami. A later court decision ruled that the Jones Act, a US marine law, did not apply, because foreign crew contracts were commercial in nature. An independent arbitrator in the ship's flag state or the worker's country of origin had to resolve issues between the crew and the company. Whether or not a cruise-ship victim gets justice is sometimes dependent on the ship's position at the time the crime occurred, such as the 2012 acquittal of a South African man accused of a rape that apparently happened while the ship was in international waters.

In 2011, Carnival reflagged all the Cunard ships from Southampton, England, to Hamilton, Bermuda. It was only a matter of time before Carnival bought Cunard a new flag from one of the lax open registries; the Union Jack was actually a holdout in an industry that now relies on what are called flags of convenience. Flags of convenience originated during Prohibition in the 1920s, when American ocean liners re-flagged their ships to Panama so their passengers could continue boozing up at sea.

A Cunard press release justified the move as a step into the wedding-at-sea market that the Union Jack forbids because the United Kingdom restricts legal weddings to churches, registry offices, and other approved venues, as Cruise Critic and other cruise-coverage sites reported. But it would be naive to overlook the savings that Carnival stood to gain from ditching the British flag. With a Bermuda registration, Cunard ships are not subject to British labor laws. Britain, for example, introduced the 2010 Equality Act, which legislated that all European Union employees receive the same wage as British workers. (Britain, at least at time of writing, was still in the EU.) According to the cruise review site Cruise Critic, Cunard employs a sizeable Eastern European crew for the dining room and bar, and "its increased wage bill no doubt played a part in this decision." Thanks to its Bermuda flag, that law no longer applies to the *Queen Mary 2*. The Bermuda registration also frees Cunard ships from other inconvenient legislation and places the investigation of any shipboard crimes in the hands of Bermudan authorities. Legally, Cunard's fleet was no longer based in the United Kingdom—and die-hard Cunardists left angry messages on Cunard's Facebook page following the switch—but because Bermuda is a British Overseas Territory, the ships still fly the United Kingdom's Red Ensign flag. That flag has no doubt fooled many an unassuming Cunard passenger looking for the authentic British experience.

Of the 265 cruise ships in the world, the majority are registered in the Bahamas, followed by Malta and Panama, according to a 2015 database created by the Columbia Journalism School. These countries ask few questions, impose little regulation, and offer generous free-market latitude, an issue that the International Transport Workers' Federation and other academics have covered for years. They're the offshore banking of ocean travel, and in fact the Tax Justice Network identifies many of the popular flags-of-convenience countries as tax havens.

Policing the seas is an expensive, perhaps impossible, job. Ship flags are a way of assigning responsibility, an attempt to impose the rule of law on a vast, challenging territory. Flags of convenience have muddied these waters forever. They're not only a problem for cruise ships; they're the deep-rooted cause behind so much lawlessness at sea: ocean pollution, dumping, overfishing, and slavery. Shadowy ship owners can duck fines for oil spills or hide atrocious working conditions on board. They can mask the amount of illegally caught fish they bring in. When passengers and crew board a cruise ship, they might not understand that neither the laws of their country nor of the country where the company is based apply when something goes wrong. They're a maze to navigate afterward. Flags of convenience can make the ocean an ideal place to commit a crime.

Lawyer Jim Walker knows a thing or two about crimes at sea. For many years he worked as the trial lawyer for the now-defunct Dolphin and Majesty cruise ships. Now he specializes in bringing cases against the cruise-ship industry. His bread and butter is sexual assaults and workplace injuries: women and children groped and even raped, often by staff, and waiters with carpal tunnel syndrome from carrying loaded trays across a yawing dining room. His blog, Cruise Law News, allows him to bring attention to all the unprovable cases he couldn't pursue: the disappearances, the overboards, the drug busts. Murders at sea are particularly difficult to prove, but Walker suspects that about a quarter of deaths on cruise ships involve some level of foul play. He has his own suspicions about certain cases, like the disappearance of a pretty crew member on a Mediterranean ship, the chip mysteriously missing from her cellphone. His hunches are based not only on his legal experience but also the many years he spent working for the industry. "There are so many incredible stories that happen at sea," Walker observes. "There're a lot of things I saw on the other side."

Historically, cruise ships were not required to report disappearances publicly, but each company gathered and reported casualties to their ships' flag state. The 2010 Cruise Vessel Security and Safety Act mandated that American deaths on ships operating out of US ports must be submitted to the Federal Bureau of Investigation, but, Walker notes, they still have little incentive to report something that might damage the corporate brand. Sometimes they delay reporting a death, so social media becomes a typical way families first hear the bad news. Walker is part of a two-person tag team that acts as an unofficial watchdog on the industry. His other half is the sociologist Ross Klein. Back in 2007, Walker and Klein both attended a meeting between the major cruise lines and the lobbying group International Cruise Victims, founded by a father whose daughter went missing on a cruise ship. Klein remembers that the cruise-line representatives told them that they would not report overboards unless they received an inquiry from the media. Klein pointed out an obvious flaw with that system. "How is the media going to make an inquiry if they don't know about it?" he asked them. To me he reported, "They said, well, that's not their problem."

Crimes ought to be reported to the ship's country of registration. But, Walker explains, it would be almost impossible to mount a criminal case after the ship reached land. All the evidence, including the witnesses, become much harder to track down. When an incident is reported, a single officer flies in for what is a one-day investigation at best. "You never have Bermuda or Panama, certainly not the Bahamas, coming aboard and conducting a real investigation," Walker says, adding that most flag states never send a representative to the scene of a crime on a foreign-flagged ship. It's like a small town with no independent law enforcement for hundreds of miles in all directions. If the police took statements, preserved evidence, and photographed the crime scene, Walker notes, all of that evidence would be discoverable and could be obtained by counsel. It's different with a cruise-ship investigation. "Any investigation that the ship is doing is basically done in anticipation of defending a lawsuit," he explains, "and it's privileged."

Ross Klein maintains a website (www.CruiseJunkie.com) where he keeps a grim tally of all the crimes, misdemeanors, and mishaps that happen on board cruise ships, year after year after year: murders, rapes, robberies, drug busts, environmental crimes, groundings, sinkings, and more. One of the most infamous cruise-ship crimes of recent years happened on board the *Queen Mary 2*. From 2007 to 2011, an English childcare supervisor abused over a dozen boys on all three of the Cunard ships where he worked, according to BBC reporting. Rape and sexual assault account for the vast majority (70 percent) of crimes reported on cruise ships, but there's a plethora of unreported and underreported crimes. The *L.A. Times* published a bombshell story in 2007 that uncovered 273 assaults from 2003 to 2005 on board Royal Caribbean ships, the second-largest company after Carnival. Royal Caribbean, it is worth noting, reported only sixty-six sexual assaults on its ships in that period. In 2010, the United States passed the Cruise Vessel Security and Safety Act in an effort to tamp down on underreporting. The law requires passenger ships cruising in and out of US ports to report crimes involving American citizens, but only specific incidents, to the Federal Bureau of Investigation. For instance, cruise ships are only required to report thefts to the FBI when more than $10,000 is stolen—an unlikely sum for passengers or crew to carry on board. No wonder, then, that of all passengers on all the cruise ships, only twenty-two thefts occurred on board in 2018, at least according to the industry-wide numbers reported to the FBI. Other loopholes, too, suggest more crime is going on at sea than what surfaces in official reports. California congresswoman Doris Matsui, who cosponsored the 2010 Cruise Vessel Security and Safety Act, has introduced new legislation to strengthen security and reporting on board.

Man-overboards, or MOBs as they're called in the industry, are more common than you might think. Every year, between fifteen and twenty-five people fall off a cruise ship, according to sociologist Ross Klein's ongoing tally on his website CruiseJunkie.com. This works out to an average of one overboard death every two weeks. (This total doesn't include the deaths that occur inside the ship or during onshore excursions.) There are occasional survival stories. After falling off Royal Caribbean's *Oasis of the Sea* near the coast of Cozumel, Mexico, a 22-year-old man bobbed for five hours before a passing Disney cruise ship spotted and saved him. (He later told the port authority that he had no idea how he fell off.) A fall from the ship's decks down onto hard ocean could snap a neck, but it's also likely the person will drown or, especially in the case of the North Atlantic, succumb to hypothermia. Perhaps it's expected that the occasional passenger, drunk and partying, might fall over the railings. But ship staff, who know the vessel and its rules better than they do their own home, are more likely to fall off. In 2016, cruise-ship workers made up just over half of all the overboard incidents counted by Ross Klein on CruiseJunkie.com, even though the majority of cruise ships has more passengers than crew on board. At any given time, the *Queen Mary 2* is sailing with twice as many passengers as crew.

Some ships have automatic man-overboard systems that alert the bridge when something (or someone) goes over the rails. Why isn't this technology a standard feature on cruise ships? Jim Walker says that the cruise lines argue that there are too many false alerts, like seabirds or cocktail glasses, to make them worthwhile. Collecting and analyzing the data around overboard deaths would be a step in the right direction, Ross Klein said. The majority of overboard deaths happen between midnight and 4:00 A.M. With information like that, Klein said, cruise ship companies could at least take some preventative steps.

❧

Two days after Favio disappeared, the *Queen Mary 2* docked in Halifax, Nova Scotia, Canada, only five hours behind schedule. Halifax police officers and the Royal Canadian Mounted Police went to meet the ship and interview the crew about Favio. So did the media. One unnamed crew member told the British newspaper the *Daily Telegraph* that Favio had been crying and screaming that he wanted to throw himself off the ship. That staff member also said that the crew believed Favio had gotten into a fight with another coworker. Captain Oprey, the *Telegraph* reported, banned the crew from drinking alcohol on the ship and removed their deck privileges, meaning access to passenger decks, including the one from which Favio fell. A similar story ran in the Halifax newspaper the *Chronicle Herald*, which ended with praise for the *Queen Mary 2*'s dance venue, planetarium, and 3D cinema. Cruise-ship traffic is a valuable source of revenue for the ports where the ships dock, and the cruising industry spends millions each year on newspaper advertising. No one wants to rock the boat.

According to Ross Klein, victim blaming is standard practice after a man-overboard case, and few media outlets challenge that. Reports of incidents on board remain industry friendly. "Seventy to eighty percent of the time when you read about someone going overboard, it's always they fell or they jumped," he said. "It's conclusively stated."

After the Halifax authorities interviewed the crew and captain, they turned their report over to the ship's flag state, Bermuda. The *Queen Mary 2* departed Halifax and headed for New York City, where an investigator from Bermuda's Department of Maritime Administration would meet the ship. By the time the investigator arrived, however, some witnesses had already left. According to the *Telegraph*, four crew members quit and flew home.

Six months after Favio died, the *Queen Mary 2* pulled into the port of Valparaiso, Chile, as part of the ship's tenth-anniversary tour. Valparaiso is only two hundred kilometers from Favio's home in Rancagua. At one point, he and Santiago had imagined showing their families and friends around the boat and sharing crazy stories of their time travelling the world. Then, they said, they would walk back onto Chilean soil and leave the ship behind forever.

Favio was gone, but on that bright February morning, his face was everywhere. A group of protesters unfurled a blue banner that read JUSTICIA PARA FAVIO in hand-painted letters. They waved photos of him dressed in his kitchen whites. They handed out pamphlets in two languages. TV reporters crowded around his mother and sister as they told their story: they wanted the CCTV tapes on board, they wanted Favio's insurance documents, they wanted answers to the most fundamental questions about what happened to Favio. His family had pressured the Chilean police to search the ship when it arrived. As Favio's friends waved signs and talked to cruise-ship passengers outside *Queen Mary 2*, detectives from the Chilean police went up the gangway to interview the captain and crew and collect the footage from the night Favio went missing.

This was not Carolina's first attempt to get answers from Cunard. A month after Favio's disappearance, Cunard flew his family across the Atlantic to Southampton, where Cunard is headquartered at Carnival House. For that whole month, Carolina had felt like she was in the dark about the investigation. She'd waded through the English-language news stories and tried to piece together what had happened to her son. On Facebook, Favio's friends were flooding her inbox with stories. One friend told her that Favio and his Filipino coworkers had had a fight two weeks before he went overboard. According to that story, the chefs wanted to tattoo Favio—a way of marking him as one of their own—but Favio rebuffed the invitation into the group. This insulted the Filipinos and led to an argument. She heard another story about a scuffle in a bathroom: one crew member had held Favio down and injected him with some unknown drug. Someone else told her he'd gotten into an argument over a woman, while others dismissed the story as rumor. Favio's girlfriend assured Carolina that she and Favio hadn't fought that night, no matter what the newspapers reported; they were happy. Then the messages stopped abruptly. When Carolina sent messages of her own to ask why no one would talk to her anymore, she heard that someone was making life difficult for the crew on board. Later on, she heard Favio's friends and crew members on board attributed the crackdown to the ship's captain and security. But in England, she planned to make the investigators answer all her questions in person.

❧

Carolina, along with Favio's sister and father, went to England to meet with Cunard's representatives at the Southampton Holiday Inn where they stayed. According to the family, Cunard's head of captains, a Cunard human resources representative, and a security manager met with the family. Sebastian Lorenzini, the Chilean consul to the United Kingdom, translated the meeting. Throughout the meeting, Carolina felt that Cunard was pressuring Lorenzini. The company also presented her with a diagram of Favio's last steps on board and played two short video clips of the moments directly before his fall. They told her that the investigation had concluded that Favio had committed suicide. His death certificate, stamped by the Registry General of Bermuda, gave "suicide" as the cause of death.

Carolina had heard from Favio's friends on board that two crew members left the ship after it docked in New York City. She tried to ask the investigators if these people were with Favio the night he disappeared. Were they interviewed before they left the ship? If so, what did they say, and why did they leave the ship? According to Carolina, Cunard's representatives told her they couldn't release the names or nationalities of anyone involved in the investigation. She said that when she asked questions about the stories that Favio's friends had shared with her—like the fight with his coworkers and the bathroom scuffle—the staff only wanted to know who gave her that information. They couldn't—or wouldn't—confirm if they had investigated these incidents.

Carolina flew home to Chile with more questions than when she had left. Five months after that meeting, the Chilean state department forwarded the family a copy of Bermuda's investigation into Favio's death, translated by the government. In her opinion, the investigation painted Favio as a drug addict who snorted cocaine on land and smoked synthetic marijuana at sea. The document described the events leading up to the fall: Favio drinking in the crew bar, smoking on the crew deck, and suddenly acting erratically after he emerged from a washroom, and his coworkers trying and failing to take him back to his cabin. But Cunard, she said, never showed the family the complete surveillance footage of Favio's last night, even though it would have helped them understand the series of events described in Bermuda's official report. According to Carolina,

Cunard told her in the face-to-face meeting that Favio's fall was a suicide but later concluded that his suicide was actually brought on by drug abuse.

This explanation didn't satisfy Carolina. Cunard has a zero-tolerance drug policy, and Favio, like all Cunard crew members, took drug tests twice every six-month contract. His came back clean every time. Cunard had no body to test, so the allegations about Favio's drug habits were hearsay at best and unverifiable by an independent third party. She created a Facebook group (VAMOS FAVIO . . . ¡TE ESPERAMOS!) that acquired thousands of members. She started talking to Chilean newspapers and any other journalist interested in what had happened to her son. She went on national TV. And she continued to seek answers, including consulting a fortune-teller. Now she stood on the Valparaiso quay, surrounded by supporters, the *Queen Mary 2* towering above her.

Santiago Gutiérrez also walked along the quay. He kept his head down, his eyes averted. He didn't make eye contact with Carolina as she spoke to television reporters. He avoided the crush of tourists, the media, and the protesters waving signs of his friend's face. Santiago no longer partnered with Favio's family in trying to figure out what had happened. Once, he had connected his ship friends with Carolina and passed along important information about how the ship operated and government contacts to ask for help. But after a reporter published a story in which Santiago suggested that Favio might have been depressed and committed suicide, Santiago's relationship with Favio's family and friends soured. His Facebook inbox filled with angry messages that accused him of selling out Favio—so many messages that he stopped reading and shut down his account. He was still cordial with Carolina, but he'd stopped talking with the rest of Favio's circle. Santiago led his girlfriend along a circuitous route and boarded the gangway of the *Queen Mary 2*.

Inside, the ship swarmed with Chilean investigators, taking pictures and measurements of the deck where Favio fell and seizing videotapes from the CCTV system. Captain Kevin Oprey walked past Santiago, talking hurriedly with Chilean officials. Favio's former cabin was cordoned off. Some of the ship's crew were kept on board for questioning, while most of the passengers had gone on land to explore the city, giving Santiago a chance to see his old friends and coworkers during a rare moment of downtime on board the ship. He could barely walk across a deck or down a corridor without running into someone he knew. "Most people are glad to grab your hand and say they're happy for you [and your new life]," he remembered.

Eventually Santiago and his girlfriend cut away from the crowded decks to take a tour of the ship by themselves. They entered a three-story art deco dining room: the Britannia, the largest dining room on board, with a sweeping staircase and a vintage tapestry of an ocean liner. The Britannia could serve around 1,300 people at once. This is where Santiago dreamed of working when he first arrived on *Queen Mary 2*. On his third contract, he got there, promoted from monitoring hot plates in the King's Court to overseeing the pasta section in Britannia.

By late afternoon, the Valparaiso port had grown quiet and deserted. Carolina and the rest of Favio's family and friends had finished protesting in front of the *Queen Mary 2* and moved the demonstration to a downtown square in Valparaiso. *Chinchineros*, traditional Chilean street performers, danced in front of the blue JUSTICIA PARA FAVIO banner, pounding the drums strapped to their backs. Online, friends and family across the country posted pictures of themselves holding the same sign. When Santiago and his girlfriend walked down the gangway, he later said, "Nobody was outside, no more protesters, no more journalists, no more nothing." He and his girlfriend got into their car and drove home.

Some of the people Santiago had spoken to on board had talked about Favio. Long before he went overboard, they told Santiago, Favio changed. He stopped caring. He showed up late to work. He was drinking and playing music late into the night at the crew bar. Twice the executive chef called Favio into his office. The other chefs on the line assumed he would be fired. Instead he walked out reprimanded but still on the line. The problems continued. (In 2019, three years after Santiago told me about his visit to the *Queen Mary 2* in Valparaiso, I called to confirm what he heard on board that day. He did not remember the crew's stories about Favio's flagging performance before he disappeared.)

"When I was living there, I tried to take care of him," Santiago said. He couldn't help but feel like he might have stopped whatever happened to Favio that night.

But could he? According to a Sailors' Society survey of one thousand seafarers, a quarter reported feeling depressed over a two-week period, and nearly half of those did not ask for help. Even with support services on board, talking about private struggles remains taboo in the work-hard, play-hard culture at sea. In certain countries, it's hard enough just to snag an interview with a cruise ship, and thousands of other candidates are willing to work, so crew members are reluctant to tell their superiors when they're struggling to cope with the reality of cruise-ship life. The pressure, work, and loneliness on board might particularly affect South American workers. The Mission to Seafarers, a sea worker advocacy group, has devised a happiness index. South Americans rank near the bottom.

"People break every day," says Brian David Bruns, the author of *Cruise Confidential*, a memoir of his years working on cruise ships. "They get mentally and emotionally crushed by this job. And a lot of people can't quit. Their families are depending on them. They need the money. I have seen so much of that." According to Bruns, the routine breaks everyone down in the end. Some staff crash after a few days, others a few years. Rarely does someone make it through a lifetime of work.

The first year Santiago worked on the *Queen Mary 2*, he remembered, the ship held a crew party in the luggage area. The free drinks flowed. "You're trying to fix the environment with alcohol. What do you expect?" he asked. When he started his second contract, he pulled back from the partying he had done in his first term because he had been demoted to a behind-the-scenes position prepping vegetables. But he still wanted to show his superiors that he could handle the job, that he could be a machine. The promotion to his new position in the Britannia didn't leave him any time for side projects. He gave up drumming in an on-board band with a Peruvian, a Filipino, two English women, and a Russian. (They called the band Black Star, a little dig at Cunard's illustrious White Star service.) His responsibilities increased, and so did the stress and infighting in the kitchen. Sometimes after working ten or eleven hours, he would spend his last waking hours cleaning his cabin before inspection the next day. By his second and third trips around the world, Santiago gave up shore visits. No matter how exotic or far-flung the destinations, he chose to sleep in his cabin instead of pounding the pavement and spending money in cities he had already seen.

As the contracts rolled into each other, the months turning into years, Santiago felt himself disconnecting from life. His long workdays seemed a never-ending cycle of stress and fatigue, bosses hectoring him from all sides to uphold Cunard's famous "White Star Service," always yelling at him to be professional, professional, professional. Meanwhile, larger questions gnawed at him. Santiago was nearing thirty. No one was waiting for him back home. When he looked around the kitchen, he saw chefs with families they rarely saw. If he started a family in Chile, he would be forced to leave them for six months at a time. But quitting wasn't an easy option. "Money is money," he said later, "but I could feel my life getting worse."

The *Queen Mary 2* gave Favio a job, security, professional advancement, and money that he couldn't make in Chile. But although Favio had paid off his student debt, as he planned, he still wasn't debt free. According to Carolina, Favio had started saving to buy an apartment in Chile, but Santiago remembered that he often returned to the ship short on cash after his vacations. Because breaks between contracts are unpaid, this wouldn't be uncommon among ship staff.

When I was sixteen, I unwillingly went on a Mediterranean cruise with my equally unenthusiastic sister and our mother, who had just read *Captain Corelli's Mandolin* and felt an intense urge to travel in the Greek islands. I found the mood on the ship jarringly happy. The crew members seemed too ecstatic, too eager to laugh at a guest's off-color joke or answer some mundane question they must have heard a hundred times before. One incident in particular stayed with me. During one of our last dinners on board, a waiter arrived at our table with a tray of Baked Alaska in hand. "Opa," he said listlessly, slamming the dessert down on our table and lighting the meringue on fire. He moved to the next table, and the next, and the next, grimly repeating the routine each time.

Favio's situation on board was not unique. He drank, he partied, he worked constantly, and still he couldn't save money fast enough to get out of the cycle of cruise-ship contracts. Walking the decks alone that night, upset, burned out, Favio had no safety net, literal or metaphorical, to catch him. No alarm signalled his fall. In the developed, postindustrial world, crossing the ocean can be whatever we want it to be: a family-friendly vacation, a party cruise, a trip to distant lands. As more people in the developing world enter the middle class and shift their view of the ocean from industrial to recreation, cruise-ship profits will soar. Business is booming in emerging markets like China; the industry estimates that it will host four million annual cruisers from that country by 2020. Meanwhile, for people in low-wage countries, a cruise-ship job is still a good gig compared to working on a fishing boat or a container ship. There is a long supply of both types of people—the ocean lover and the ocean worker, the served and the server—and everyone is in the same boat together.

Favio's family continues to push for more information about what happened on the night he disappeared. They petitioned the Chilean president for help and consulted with lawyers. They haven't received any financial compensation from Cunard, although Carolina says the company called to offer the family twenty-five million Chilean pesos ($39,000 at the time) on the condition that they would never comment publicly on Favio's case again. Carolina isn't a fervent Catholic, but it's difficult for her to move on without putting her son's body to rest in the ground. The family has still not held a funeral for Favio. But the most difficult part of moving on is not knowing what happened to her son that night. Maybe it was Favio's time to go, she says, but not this way. His family hopes that his case might raise more awareness about the reality of work on a cruise ship. The paycheck can blind workers to the demanding conditions and diminished rights they'll have on the high seas.

When she met with Cunard representatives in England, Carolina watched only a few video clips of her son's last night, compiled from the *Queen Mary 2*'s extensive surveillance system. In a series of short, disconnected clips she saw him walking the corridors, roaming the passenger-only decks. She saw him stumbling, his knuckles scraped. In these edited clips, Favio was always alone. Finally, she saw a man leaning against the railing in front of him on deck 7. Rain fell and the video was grainy, obscuring his face. She wanted to be absolutely certain it was Favio on the deck, but she wasn't. With one hand on the railing, the man leaned. He seemed like he was thinking. What was he thinking about? Carolina wondered. And then his body went limp, as though he had fallen asleep. The video lasted three or four minutes as the man stood at the edge, the dark ocean beyond. If someone in security was monitoring the surveillance cameras, they could have come to help the man. As Carolina described the experience of watching her son's last moments, I could sense how much she wanted to rewind the tape, to go to Favio's side and comfort him.

And then he fell.

In 2015, twenty-six other people fell off cruise ships. Another fifty-two died during excursions, drowned in swimming pools, or suffered medical emergencies far from land, according to sociologist Ross Klein's ongoing tally on his website CruiseJunkie.com. But the official year-end tally that Carnival submits to the US Department of Transportation showed only two people died on Carnival cruise ships that year. That number only includes the American citizens. As a Chilean, Favio's death didn't count.

SEVEN

SAVING A PREHISTORIC FISH

One afternoon not long ago, I looked a fish in the eye and saw something beautiful. Her small pupils revealed a deepening maze of tawny, glistening yellows. Her scales at first looked gray and matte, but when I moved in closer—and I mean nose-to-scale close—her armored skin became a kaleidoscopic swirl of colors and shapes: purple diamonds, coral glints, blots of black like spilled ink. On each side of her half-moon gills was a starburst of metallic purples and pinks, like she was wearing the most awesome earrings in the world. If it hadn't been a rainy October day, the quicksilver colors of her skin would have caught the sun.

Then, with a quick flash of her tail, the sturgeon and her hallucinogenic-colored skin and eyes were gone, released back into the water. Every second she spent with me out of the estuary jeopardized her return to the water. Here in the Fraser River, she and thousands of other white sturgeon like her swim in and out of the Pacific Ocean. It's where Erin Stoddard, a stocky 53-year-old fisheries biologist, has studied this mysterious fish for fifteen years.

The modern world tends to leapfrog from crisis to crisis, headline to headline. Some stories about the ocean make front-page news, like the Great Pacific Garbage Patch or refugees in the Mediterranean. Very rarely does the quiet demise of an uncharismatic mud-trawling fish like the white sturgeon make headlines—unless, of course, that fish goes extinct, which is not impossible today.

Most would not call the sturgeon beautiful. It certainly doesn't have the aesthetic appeal of a frolicking dolphin or the cuddliness of a big-eyed seal. Instead, it has the plated skin of an alligator, the upright back tail and white belly of a shark, and the beady eyes of a snake. Sturgeon are widely distributed across the northern hemisphere, from rivers in China and Russia to the Great Lakes and the Sacramento River. One of the largest sturgeon ever reeled in on the Fraser River in recent years stretched over twelve feet long, rivalling the length of the boat that caught it, and weighing more than a grand piano. A sturgeon trawls the river bottom, its catfish-like whiskers dragging through the mud, its pointed chin pockmarked with papillar taste buds. It looks like a relic from a fiercer time, perhaps as early as the Carboniferous period according to new research, and the sturgeon's anatomy has changed remarkably little since it evolved in prehistoric oceans and rivers. This ancient fish has survived, exactly as it is today, for millions of years, not to mention the last six thousand years of modern human history. To fish nerds, this is a cool fish.

Nearly every population of sturgeon around the world is in some sort of trouble. In the mega-dammed, overdeveloped Yangtze River, the wild Chinese sturgeon will likely go extinct before 2030, scientists predict. The once abundant Alabama sturgeon is now so rare it's gained an almost mythic status in that state. In the Caspian Sea, people have poached and overfished the massive beluga sturgeon to near extinction. In the Lower Fraser River, where Erin Stoddard worked, accumulating and piecing together patchy information about the white sturgeon, this fish swims through some of the most dammed, dyked, and developed land in the Pacific Northwest. New ports, new bridges, new dykes, new levies, and new drainage projects are always in the works right in the sturgeon's habitat. The rush to develop shows no sign of stopping. The International Union for Conservation of Nature has estimated that the sturgeon group of species is more endangered than any other in the world, with 85 percent at risk of extinction. Aquaculture may be the only future left for sturgeon: already 90 percent of young beluga sturgeon come from artificial breeding.

At first glance the white sturgeon population in the Lower Fraser River seems healthy. In 2015, around 45,000 sturgeon swam and spawned and fed in the river and out in the Pacific Ocean. This seems like a high number, but it actually hides a worrying reality: sturgeon can live over a hundred years, and many of those fish are on the older side.

"That's part of the difficulty of being a sturgeon biologist," said Erin Stoddard. He is directly responsible for the future of white sturgeon in these parts. "Their generations are similar to ours. How can you tell when you've come to a recovery state, or when it's time to do other management, if it takes thirty years to figure that out?"

Along the Fraser River, each new condo development and overpass takes another bite out of the naturally loose gravel along the riverbanks, where sturgeon hide their eggs. Studies show that the next generation is struggling to reach sexual maturity: since 2004, juvenile numbers have declined as much as 50 percent. Nearby populations in the Columbia, Kootenay, and Nechako Rivers are in worse condition. For years now, they haven't had a significant spawn that would boost their numbers. This is particularly worrying in a late-blooming fish like the sturgeon. Males start spawning around fifteen years old and later, while females can take eighteen years years to produce eggs. By the time the next generation arrives, the population may have already suffered a catastrophic, ecosystem-destabilizing decline.

At least we now have a name for this pattern of slow and steady demise of species like the sturgeon: the Anthropocene, which has ushered in the world's sixth mass extinction. Over the last few environmentally turbulent years, the Anthropocene has moved from scientific discussion to the cover of the *Economist.* Scientists debate exactly when the Anthropocene began, but most agree we're already living in it. Plastic pollution is one hallmark of this scary new era. Another is species extinction. Aquatic habitats are threatened from multiple angles: pollution, ocean acidification and warming, coastal development and erosion, invasive species, poaching and overfishing. This time, extinction will not be a matter of chance asteroids but the cumulative actions of one species: our own.

~

It was a good day to be Erin Stoddard. Normally, he spent his days inside a drab government building in a suburb of Vancouver, managing and studying the fish populations of the Lower Fraser River from his computer. In terms of sturgeon, that might mean reprimanding fishing guides for incorrectly handling the fish, or helping his team crack down on restaurants with poached sturgeon swimming in their tanks, and even closing the recreational fishery down altogether, which he considered doing for the first time in 2015 when the river temperatures rose to unprecedented levels and the water level dropped so drastically. But this foggy morning, he was loading up his research boat for one of those rare days in the field, catching sturgeon and tagging them with acoustic trackers. It was year three of Erin's ten-year study tracking and gathering basic information about white sturgeon. A dozen boats sat nearby on the gravel shore, waiting for the annual sturgeon-fishing derby to begin.

Here are some things we know and don't know about sturgeon. Sturgeon are not just freshwater fishes. They spawn in rivers, but they roam out to sea regularly. Like salmon or eelfish, they're a group of fish that moves between inland water and tidal water—a piscine scorecard for how humans are treating both habitats. We know sturgeon swim out to the ocean, but we don't know where they go or what they do out there. In 2009, sturgeon flummoxed researchers when 1,500 fish stranded themselves on the banks of Washington State's Puget Sound. Was this a feeding frenzy gone wrong when the tide lowered? Because scientists are still unclear on what exactly sturgeon do in the ocean, it's hard to know.

Sturgeon like to jump, but we don't know why. This may sound like normal fish behavior, but a sturgeon jump is like a rocket liftoff compared to the ripple-like effects of other jumping fish. They shoot to the surface with such force and speed that they launch themselves out and over the water, twisting and turning and smacking down on the surface—sometimes with terrible results. A sturgeon once leapt into a boat in Florida and killed a five-year-old girl. White sturgeon get jumpy in February. Are they feeding? Communicating? We don't know.

Another mystery is the massive sturgeon ball. In the winter of 2008, a sonar survey revealed a big pile of something at the bottom of the sixty-meter-high Bonneville Dam in Washington State. State officials figured the dam must have been crumbling into the Columbia River. Instead they discovered a gigantic knot of sixty thousand sturgeon congregating fifteen meters deep through the water. That one cluster accounted for up to 10 percent of the entire Lower Columbia River population. Experts said it might have been a defensive tactic against predatory sea lions. Then again, maybe they were just feeling social.

"It would be great if even one fish went out to the ocean and came back," Erin said about the more than 140 fish he's tracking in his study. That fish could give him an idea of why sturgeon go to the ocean, what time of year they leave, how much time they spend there, and what sex and size might mean for their sea mission. When the fish returned to the river to spawn, he could gather information about whether or not it went back to its usual spawning grounds. If that sturgeon's chosen spawning site had been destroyed or altered, Erin could find out whether or not it learned to adapt.

Washington State fisheries biologist Michael Parsley has compared the white sturgeon to the woolly mammoth or saber-toothed tiger: species that disappeared from Earth long before we could figure them out. We still have a chance to figure out the mysteries of sturgeon.

The sturgeon derby would help Erin add more fish to his tracking study. Normally, if Erin wanted to add more fish to his study, he had to catch them himself, and that wasn't always easy. He once spent three hours wrestling a sturgeon to the surface. When he finally raised the fish, it measured over two and a half meters. Big, but not the giant he was expecting after such an epic fight. Today he'd piggyback on the free labor of a dozen sturgeon-loving fishers. Any boat that reeled in a sturgeon longer than 160 centimeters was a new data point.

Everyone I met at the derby spoke about sturgeon in awestruck tones. Along with the volunteer fishers on the other derby boats, Erin's helpers for the day were a park ranger in training, a sturgeon biologist, and Greg George, Erin's wiry, excitable colleague at the Ministry of Forests, Lands, National Resource Operations and Rural Development. Erin will take on his research boat pretty much anyone who doesn't mind bobbing on the water all day long. Over the coming days I would spend with Erin tagging sturgeon, I would be his scrub nurse, passing him scalpels and gauze during surgery; his data recorder, taking down the length and girth of tagged sturgeon; and his personal assistant. "Where's my wallet, Laura?" he'd ask me.

All of us loaded fishing gear and coolers onto the boat, along with waders, life jackets, toolboxes, and water bottles. "Don't drop that on the ground," Erin said after I accidentally dragged an underwater microphone on the floor. He delivered the order with such a cartoonish wince that I didn't mind the scolding and laughed. For someone who has to deal with all types of people on his boat—paid and unpaid, trained and untrained, biologists, fishers, fish enthusiasts, students, and writers—he has the necessary touch for fun straight talk. When the park ranger and the biologist on board couldn't figure out how to close Erin's pocket knife, he observed, "You guys aren't too mechanically inclined," dragging out *me-cha-ni-cal-ly* as he folded the knife shut.

Unfortunately, I wasn't much help in the day's goal of reeling up sturgeon. I've caught a few fish in my life but usually by accident, not intention, and never one as large and powerful as a sturgeon. Erin grew up in the interior of British Columbia, fishing from the age of four. He knows how to set the hook deep into the sturgeon's bloated lips and to reel in aggressively, guiding the fish up toward the surface and not giving it a moment's slack until it's on board.

Until the 1980s, no one tracked sturgeon here, and no one cared much about them. They were largely considered a junk fish, except by the Indigenous peoples who have lived along these coastlines and rivers for over ten thousand years. According to Stó:lō oral tradition, the sturgeon was first created when a young woman was cast into the river. The fish resurfaces in Stó:lō contemporary artwork, too, a link between past, present, and future generations.

In the 1990s, dozens of dead, bloated female sturgeon washed up mysteriously on the banks of the Fraser. There was no clear explanation for why they died. For many people in the region, this was likely the first time they heard that a huge fish, dating back to the time of the dinosaurs, was swimming in the river behind their backyards. The provincial and federal government banned the killing of sturgeon in the Lower Fraser and created a catch-and-release fishery in its place. Even among fishers who know and catch sturgeon, Erin still has to bust the myth that sturgeon are sluggish bottom-feeders who go into low-energy hibernation at the same chosen site year after year. Not true, said Erin. During the winters, he's witnessed sturgeon actively hunt and snap at his bait. They can outrun spawning salmon. They can travel as much as fifty kilometers and migrate into the sea, where they cover hundreds of kilometres along the Pacific Northwest coast.

With 140 sturgeon in the waterway with telemetric radio devices in their bellies, Erin's tracking study will be the first on a population of wild white sturgeon swimming through an undammed river. With another twenty-three sturgeon tagged, he hoped to answer why the next generation was failing to thrive. He also hoped to protect their spawning banks. With a reasonable level of confirmed use of a certain riverbank, Erin could take steps to protect that site from the squeeze of development.

An apex predator like the sturgeon can have tremendous sway over the ecosystem. They eat everything from the smallest water bugs to the migrating chinook salmon. They feed the river, too, with their eggs and larvae. The sturgeon's decline will reverberate down through the levels of the ecosystem, upsetting the already delicate balance of life. When the future of an ancient, long-lived animal is on the line, we simply don't know all the connections we might be losing when it disappears. Like every scientist, Erin was up against time and technology. The fish he was trying to save were years from being born. Research done today could help shape the forces that were reshaping the water they would live in one day.

At 11:00 A.M., the derby boats reversed off the beach like an armada and scattered up and down the river, staking out fishing holes. Erin revved the engine on the research boat, while Greg passed out yellow earmuffs to everyone on board. The boat was a bit of a joke amongst the team. It was functional, but really, really loud. Out on the water, Erin cut the motor near a boat with a team calling itself Sturgeon Urgin'. "Hey, you got earmuffs for everyone else on the river?" the angling guide yelled over. Erin was about to retort when a flurry of action on the stern caught his attention. A young man in a ball cap was pulling on a rod bent in half in his shaking hands. Erin swivelled the boat around for a closer look as the man wrestled with something beneath the surface.

"Look at him. Guy's not even reeling in," Erin complained to Greg, standing next to him. "We're gonna be here forever." Greg cupped his hands around his mouth, like a megaphone. "Get someone else on the rod!" he heckled.

Just then a long gray flank of fish broke the surface, water splattering. The sturgeon was big, definitely an adult, and weighed over fifty pounds: exactly the kind of fish Erin needed to tag for his study. The jump was useful in another way. It meant the fish would tire quicker and make for an easier catch.

Part of the reason sport fishermen like to catch sturgeon so much is that they fight back long and hard. They're strong swimmers and they know how to throw their weight around. These fish can be the size of a small motorcycle, and some are so heavy that even hoisting them out of the water can damage their internal organs simply from the force of gravity. Many Fraser River sturgeon are caught multiple times in their lives. Some are so well known they're given nicknames, like Pig Nose, a legendary seven-hundred-pound, bulbous-nosed sturgeon who was landed for the first time in 2016 and again in 2017. It's easy to imagine that fighting a fish for three hours and lifting it out of the water has some impact on the sturgeon. Sometimes when Erin went to tag fish, he brought along PhD student Montana McLean, who studies the stress that sturgeon experience during catch-and-release fishing. When they land a fish, she takes a bit of blood.

The whole catch-and-release practice is based on the premise that fish survive after they're released to the water. In one paper, McLean and her coauthors noted that little scientific evidence exists to prove that this is true. Studies on the catch-and-release of shark and salmon show that stress, injury, and even death can happen after a fish is let go. McLean and her colleagues once took sturgeon out of the water and placed them in a sling to simulate conditions a sturgeon experiences during angling. Some fish died, while others were so tired and disoriented from the experience that they could become easy prey for predators. The concern for a female sturgeon is that the fish might become so stressed during the fight that she reabsorbs the roe into her body, as fisheries expert Marvin Rosenau told the *Vancouver Sun* after the study's release. This could explain why, over the last decade, the number of young sturgeon in the river has declined.

Although nearby populations are already classified as endangered, the Lower Fraser white sturgeon are only recommended for a "threatened" status by the Committee on the Status of Endangered Wildlife in Canada. This independent body composed of scientists and wildlife experts makes scientific recommendations to the federal government on listing animals for federal protection and research. The only reason the white sturgeon population in the Lower Fraser River is not labelled endangered, the committee noted, is their socioeconomic importance to recreational and Indigenous fishers—something the federal government does consider before listing an animal, but COSEWIC does not. The white sturgeon are likely in just as much trouble as the endangered populations farther north, but an endangered rating would shut down the lucrative sturgeon-fishing industry.

The Sturgeon Urgin' fishers traded the rod back and forth, giving everyone a chance with the first catch of the day. Meanwhile, the angling guide called out directions and maneuvered the boat to the shoreline. They would have to land a fish this big on the beach. Erin quickly pulled waders up over his cargo pants, revved the engine again, and followed the Sturgeon Urgin' team to shore. Greg and others on the research boat readied the surgical equipment, working quickly. No one wanted to delay this fish longer than necessary.

As the research boat hit the sandy banks, Erin grabbed his surgery kit and leapt into the river, water up to his waist. I followed with a clipboard and pencil. The fishers stood on the riverbank, holding the sturgeon and snapping pictures. Erin unrolled a yellow measuring tape along the fish's side. "One hundred ninety-three centimeters," he called to me, and I scrawled it down. He wrapped the tape around her girth. "Seventy-seven centimeters," he said. The fish was about the length of a single bed. It was a runt compared to the giants swimming in this river, but this fish still looked way too big for freshwater.

As we motored along the water, zipping from derby boat to derby boat, Erin pointed out small changes he'd noticed since the last time he'd travelled this way. On starboard was an inlet where a pebbly bottom broke the surface and disrupted the current. The boat rounded a corner, and he pointed out a dyke that the river had overrun.

"Have you ever heard the phrase 'death by a thousand cuts'?" Erin asked. He explained that instead of a single big impact, like a dam, he saw small impacts and wondered if eventually they would rob the river of its functionality. He genuinely did not know what it would take to make it impossible for the sturgeon to survive in the Fraser River, but it would be helpful to figure out how dire things had to get before that happened. The survival of the river was intertwined with the well-being of the sturgeon. "Rivers move like a living thing," Erin explained. "When we, the people, confine them to a small space, they tend to move more and tend to move with more force. What that does is impact habitat."

Every so often a shaved swath of forest interrupted the misty, tree-covered banks. These were log sorts, where international buyers fly in to cherry-pick the best timber. The chosen logs are sent careening down wooden skids and splash into the water. The logs leave behind rotting bark and debris that smother fish eggs and rob the water of oxygen.

The gravel-extraction sites we passed posed another serious problem for sturgeon. Gravel companies like to take the clean, naturally washed rocks from the shores and turn them into gravel for construction projects and roads. They get government money for what they present as flood protection: extracting rocks from a river that is prone to catastrophic flooding, like the 1948 flood that destroyed and damaged 2,300 homes and cost $20 million. Scientists and activists say that's a ruse. The gravel industry uses flood protection as a way to mine gravel on the government's dime, seriously damaging fish habitats in the process. Larval sturgeon need that loose gravel to hide in before they're big enough to go out and hunt.

The water itself could also generate money. Millions of hydroelectric dollars course through the waterways of British Columbia, Washington, and Oregon. Hydroelectricity is often mistaken for a benign green power, but dams can have a devastating impact on wild fish. A dam could drive up the water's temperature and oxygen levels, while restricting the river's flow: all essential parts of a fish habitat. Basically, it could turn the river into a different ecosystem: a pond. Worst of all for migrating fish like sturgeon, dams make it harder for them to access their spawning sites. In fish like salmon that spawn on schedule, a dam's impact is almost immediately apparent. If a dam stops salmon from reaching their spawning grounds in time, there will be no more salmon in the following years. With sturgeon, whose spawning patterns are not fully understood, the future is murkier. As an ancient and long-lived species, perhaps the sturgeon will figure out how to adapt. But in the Nechako and Columbia Rivers on both sides of the US–Canadian border, white sturgeon are forced to conform to dams that constrict their migrations. Neither the Nechako nor the Columbia population has fared well; both are classified as endangered.

What's impressive is that the main stem of the Fraser River, a massive waterway, still runs unchecked, from the mountainous slopes and evergreen forests inland, through dusty sagebrush and lush rain forest, out into a massive delta at the Pacific Ocean. The sturgeon population that Erin studies is one of the last in the world that still has that kind of freedom. Although the waterway is always undergoing nips and tucks from development, the fish can still migrate where they want when they need to. This is a rare opportunity to study a mystery fish's behavior in a setting as near to natural as possible.

"Those guys aren't catching any big fish, so we'll catch 'em," Erin said as he dusted the lunch crumbs off his waders. It was early afternoon and the team was huddled under the boat's awning, finishing up their packed lunches. Big fat raindrops splashed down, and the current rocked the boat from side to side. For the last hour, we'd waited and listened for a call from one of the nearby derby boats on the CB radio, hoping for someone to tell us they had a sturgeon on the line. Nothing, except radio silence.

It was a disappointing catch on a day that had required months of preparation from Erin. He still needed over twenty fish to round out the study. He opened up a plastic yogurt tub filled to the brim with fat electric-orange chum eggs and began to prepare sturgeon bait, washing his hands thoroughly before plunging them into the slimy tub. Sturgeon have a keen sense of smell, he explained, and any unnatural odor might put them off. Next he opened a toolbox that contained a stack of Sophisticated Miss–brand panty hose. He took a sticky glob of chum eggs, placed it on a pair of white nylons, cut a small circle around the roe, tied it with a topknot, and plunged a sizeable J-hook through the little bundle. I plucked a single salmon egg from the tub and popped it in my mouth, letting the oil dissolve on my tongue.

At the stern of the boat stood a row of holsters for poles. Erin slotted a fishing rod into one holster and lowered the bait into the water. A good angling guide will fish a spot for five minutes, ten at the most, and then haul anchor and move along, he said. A few minutes passed quietly. The river ran gently around the boat. Rain fell on the awning, and the wind ruffled the evergreens on shore. "We got a bite!" Greg George hollered so loud everyone jumped.

Erin gently eased the rod from the holster into his hands, keeping the line as still as possible and watching for any movement. A twitch. Something was definitely nibbling on that nylon bundle of salmon eggs. He yanked the rod high, like he was throwing it behind him, but his arms barely reached above his shoulders. We had caught our own sturgeon big enough to tag—well, Erin had, anyway.

Behind the boat, I saw a muscular fish weaving and rolling, wrapping the fishing line around its body, doing whatever it could to resist the pull to the surface. Soon the sturgeon was coasting into view, the fight slipping from its body. As it surfaced, it rotated upside down, revealing that white and somehow vulnerable belly. Erin handed the rod off to Greg and slipped his left hand inside the fish's toothless mouth. With his right hand he hoisted the whole thing out of the river. For a brief moment, Erin stood cradling the massive fish in midair, its big mouth gawping for water, before he dunked it into a water-filled operating sling set across the stern of the boat. The fish relaxed; its gills opened and it took a deep breath, revealing huge mushroom-like frills inside. Our team stood for a moment, admiring the fish.

"Okay!" Greg shouted again, snapping us out of our collective reverie. "Go, go, go! Everybody! Go!"

Scalpel, scissors, thread, gauze, antiseptic wipes, otoscope, transmitter, tape measure, clipboard, and pencil were hustled to the sturgeon's sling. Erin ran a contraption that looked like a cross between a defibrillator paddle and a grocery scanner along the fish's body. The machine beeped and Erin read a long number to Greg, who had the clipboard and was frantically sharpening the pencil with a pocketknife. "Zero, Alpha, one, three, zero, nine, three, Delta, zero, Alpha," Erin said.

This was the fish's Passive Integrated Transponder number. The PIT works the same as the gold squares on credit and debit cards that transmit money with a wave over the machine. Fish-tagging has come a long way since the 1920s, when biologists attached circular bits of celluloid stamped with serial numbers to track Fraser River salmon. According to Matthew Evenden's environmental history of the Fraser, *Fish versus Power*, they asked Indigenous groups along the river to return the tags for fifty cents a pop, but the fish tags turned into de facto currency in the Indigenous villages. Natural complications interfered with the system, as well. Tags were lost or weren't recovered until long after the spawning season.

The PIT tag is a glass-covered, copper-filled pellet small enough to fit into the needle of a syringe. The sturgeon here may look wild, but 60 percent of them already have these devices implanted in their flesh—much of it done by volunteer anglers. More advanced and less expensive tracking technology makes it possible for us to surveil increasing numbers of marine animals across the ocean. Sea turtles in Spain swim with satellites glued to their shells, and tagged great white sharks tweet their coordinates from across the sea. Close to four thousand buoys, called Argos, drift in the ocean, collecting data on surface currents, salinity, and the water's velocity, and pinging that information back to oceanographers on land.

All this ocean tracking is not only useful to scientists but also important for food security. One of the most important reasons to monitor the ocean is to catch poachers and stop overfishing and illegal fishing. But this is difficult when crimes take place in the high seas, where it's hard to keep an eye on what's happening. Nonprofits such as Global Fishing Watch use the automatic identification system data of nearly 300,000 vessels in an effort to keep tabs on fishing activities. Every day, Global Fishing Watch processes sixty million points of data through algorithms to spot unusual activity. All this information is publicly and instantly available online. But fishers from Russia and China are notorious for taking way more than their fair share from the open ocean. Out at sea, they cover their tracks by switching off their AIS ship identification system while they fish illegally or off-load catch to other boats.

Erin's study used sound to track sturgeon. When a tagged sturgeon passed one of the forty-one receivers he'd installed on the river, the receiver recorded a ping from the acoustic tag, called a VI6, which is black, metal, and about the size and weight of an AA battery. To insert the V16 tag inside the sturgeon he'd just caught, he made an incision. Dark, almost black blood pooled along the slice on the fish's belly. "This is a fat, heavy fish," Erin observed. He expected the skin to be a little thicker, but the insides revealed themselves instantly.

At first everyone referred to the fish as a boy. ("Oh, he's getting jumpy," Greg narrated nervously whenever the fish shifted in the sling.) However, it's impossible to tell sex by appearance alone. Erin had to open the fish surgically and peer in with an otoscope—those duck-billed tools that doctors use to peer inside human ears. On a mature fish like this, Erin will either see the white swollen testes of a male or the black roe of a female. Erin spotted some roe. The male was actually a female, which was even more helpful to his study. He peeled the sticker off the V16 transmitter and read its number back to Greg. This fish would be christened 22497. He nestled the transmitter inside the fish, then gently pushed it deeper into her body cavity, toward the direction of her head. Sturgeon 22497 lay quietly, apparently unaffected by a little surgery on the water. The operation lasted only a few minutes, then Erin sewed bright purple stitches into her belly.

Another part of Erin's job is performing necropsies on dead sturgeon. When he arrived on site to assess a dead sturgeon, he always worried about finding his own handiwork on the belly. So far, this hadn't happened. In fact, he has re-caught his tagged fish before and found the cut healed, the stitches invisible unless you know where to look. Once, when he found a giant dead female sturgeon, he peeled back her skin to reveal hundreds of pounds of black roe inside. This is caviar, the more famous name for sturgeon's unborn babies. It's also the reason why the wild female sturgeon will always swim with a price on her head. A fresh, fully grown white sturgeon as big as the dead female Erin found would be worth a small fortune on the black market. Not far from the river, a white sturgeon hatchery is experimenting with breeding the fish, charging $96 for thirty grams of fish eggs.

"I don't get it, but it's for the elites," Erin said about sturgeon roe, which he's tried a few times. Sometimes, during surgery, a few unfertilized salmon eggs got on his gloves and into his mouth accidentally. The taste was fishy and nothing special. He would rather eat a steak over fish roe any day.

Recently, to combat poaching, the government shut down night fishing on the water. Night was when the poachers came out on small boats with quiet outboard motors, looking to snag a giant white sturgeon. If they caught one, they might tether a larger sturgeon to the shore by her tail and keep her alive until they could find a buyer, or they'd kill the fish for its eggs and flesh.

Just as Erin was sewing the last stitches into 22497, she flailed suddenly in the sling, and water sloshed into her wound. Greg responded in an instant, mopping up the water with a paper towel, but it was too late: dirty river water had gotten in the wound. Erin sewed faster, closing the incision as fast as he could before the fish flexed again. Open-water surgeries are a balancing act. The water in the operating sling has to be deep enough for the fish to breathe but shallow enough so that a biologist can perform a sterile procedure. If the sturgeon or the boat shifted, it threw the balance off.

The operation complete, 22497 was ready to go back in the water. Erin hoisted her out of the sling and lowered her into the estuary. She was disoriented at first, barely swimming. He leaned over the boat farther, holding on to her as he pushed her deeper into the water. Now she could feel less of his hand, he explained, and more of the cold water running along her gills. A second later, her movements sharpened, as if she had woken up. Erin directed her into the current, and 22497 made a powerful snap of her fin. She vanished beneath the turbid water.

But Erin hadn't lost track of her. He lowered an underwater microphone over the side of the boat, the same microphone I had dragged on the floor at the start of the trip. Someone flicked on a set of speakers linked to the acoustic transmitters, and a crackling white noise floated over the boat. Our group sat in the rain for a moment, listening to the static crackle of the water coursing under the boat. It felt like an art installation, the sound of actual water and recorded water mixing in the air around us. Then came a series of pings, cutting through the static, like the sound of someone clicking her tongue against the roof of her mouth: one ping after another after another. Six pings confirmed a single fish detection. A number flashed across the speaker's screen: 22497 was alive and well and swimming along the bottom of the Fraser River. Now, she just needed to show Erin where she was going.

We returned to the water a few days later. Erin leaned way over the side of the boat, sweat pouring down his face. A breeze ruffled his hair, sweeping his baseball cap off his head and into the rushing current. He didn't look up, didn't even notice that his hat was gone. He had a metal hook in his hand, and he was jabbing it into the water, listening for the sound of metal on metal.

We were far from the passing swoosh of a nearby highway, in a quiet channel with little boat traffic and no rapids. The silence was deliberate. Erin had chosen this place for planting an acoustic receiver precisely because it was quiet. The acoustic receivers at the forty-one stations scattered throughout the water picked up not only the pings of passing sturgeon but also all the sounds of tidal waters rising and falling or boats passing in the estuary.

Today was about collecting the data from the acoustic network. We had already hauled up and downloaded information from two receivers today. At the third station things had gone wrong. The line tying the acoustic receiver to the shore was shorn off. Without the line, the receiver was lost beneath the surface, and just a few feet of turbid water was enough to hide its metal frame completely. The four of us—Erin, me, and his hired hands for the day—grabbed poles and hooks and started to comb the water. Each sweep brought up branches, rusty cables, and mud, but no acoustic receiver.

The $1,700 cost of the lost receiver didn't bother him; it was losing the data contained in this acoustic station. Every scientific experiment requires significant time, energy, and investment to get up and running, especially a ten-year, in-field study like Erin's. It had taken him five years to get to this point, in between all his other management duties with the government. He had spent two of those years simply waiting for year-end funds to come through and stockpiling enough V16 transmitters to implant in the sturgeon (each one costs $350). Then he went about making the study a reality, securing yet more funding, coordinating with other fish conservation groups, and working with an already installed acoustic network along the river.

This meant choosing the ideal locations for the receiver stations and overseeing their design. At the start, he spent many long days catching over a hundred sturgeon by rod and reel and surgically implanting each one with a transmitter. Each transmitter contained a lithium battery that lasted ten years, enough time for Erin to complete his study. But if he lost an acoustic receiver, all that work, money, and months of data disappeared downstream. That acoustic receiver didn't cost $1,700—it was priceless.

As the search dragged out, Erin went from chatty to stone silent. He picked up a grapnel, a nasty reverse-hooked anchor, hurled it as far as possible from the boat, and dragged it along the river bottom. It came up empty. After forty-five minutes of searching, one worker took a break, ambling along the shore looking for a discreet spot for a bathroom break. "Erin!" he called back excitedly. "Erin, I found it!" We had been searching in the wrong spot all along. The cut line we'd found was just a leftover from the receiver's last location a few years earlier. The new location was fifty meters upstream.

We pulled up the muddy receiver from the depths. Erin was ecstatic, gleeful, and finally talking again after close to an hour spent in silence. He told us that he had decided to allot just five minutes more to the search before giving up and moving on to the next station, collecting all the other receivers waiting for us downriver.

Erin took out a laptop and started to download data via Bluetooth. Numbers filled the screen, and behind those numbers lay years of work. Over the last few months, this particular station had picked up 103 sturgeon swimming past. With these numbers, he could re-create an unseen world of fish coming and going. He could put together what their movements meant and begin to understand their life stories.

During the five years he had spent planning the study, he had dreamed of answering all kinds of questions about this mysterious fish. How early in the year do sturgeon arrive at spawning sites? Are they devoted to those sites? Could they adapt to new ones? He wondered about those congregations of sturgeon that spent their winters in gigantic balls. Why do they do this? Do they move around from one site to another? Always Erin hoped that one sturgeon would leave the Fraser for the ocean and return. The ocean, however, was the end of Erin's tracking system—he had acoustic receivers as far as the estuary and no farther. Whatever sturgeon did out in the ocean would remain beyond the scope of his network. Technological advancements make tracking animal migrations in the open sea more accessible, but conducting research is still costly. Erin often collaborates with the Ocean Tracking Network, based at Dalhousie University in Halifax, Nova Scotia, whose longest acoustic monitoring line extended sixty kilometers into the Atlantic Ocean in 2015. Visiting that line for a checkup costs up to $12,000 a day in boat-chartering fees at that time. The ocean's expensive inaccessibility is why the oil and gas industry funds much of the deep-sea exploration, like the $4 million Shell Ocean Discovery XPRIZE funded by Royal Dutch Shell. Private donors, like Eric and Wendy Schmidt's research cruises and ocean institute, also fund important projects. As a provincial government biologist, Erin had to spend years saving funds and recruiting volunteer labor to spread his network as far as he has.

At 5:00 P.M., we beached the research boat and called it a day. We loaded up our gear and hitched the research boat to the pickup truck. The late autumn sun was nearly set; it had been rising when we drove to the river that morning. Twelve hours later, and the day's work was still not finished. Ahead of us lay an hour's drive back to the city and then stops at Erin's office and an aquaculture facility to drop off the gear. Another twelve hours from now, Erin would be back on the road again, driving toward the river, where he would locate the last of the acoustic receivers and collect the data they contained.

In the coming years, Erin's study would continue to unlock new information about the fish. Some of the tagged sturgeon would indeed leave for the ocean, just as Erin hoped, and he would watch in fascination as they made pointed migrations in and out of salt water at specific times of the year. "Like clockwork," he said. One spring he would catch a mature male he suspected was about to spawn at a site nearby. Later that year, he tracked that same male to the site and confirmed that, yes, that male was present during the spawn just as Erin had predicted. Another revelation: white sturgeon in the Fraser River are not loyal to just one overwintering site, as biologists have observed in other populations. Perhaps those other sturgeon populations stick to one site because they're swimming in more developed and dammed waterways than the Lower Fraser River. In short, these other populations might have adapted to fit the modern world.

Of course, the story of the white sturgeon is still unfolding. Each piece of new information unlocks more questions. "That's a standard theme with sturgeon biologists," Erin said. "The more you learn, the more you don't know." That's also a standard theme when it comes to studying the ocean. We uncover new species of deep-sea octopuses or a chain of underwater mountains, and again we realize that we simply don't know enough about the world's largest ecosystem.

Before we headed back to Vancouver, Erin stood outside the truck, deep in conversation with a man in his sixties wearing a life jacket, waders, and ball cap. They were looking at something on the man's phone. They gestured wildly and shook their heads, like they couldn't quite believe what they were seeing. The man had reeled in something no one had seen here in years, Erin said: a green sturgeon. The green sturgeon looks similar to the white, but it's an even more obscure and unknown fish. There's not enough data to say how big the population is in the Fraser River, what habitat they need, or whether they spawn here at all. The green sturgeon is not as tasty or commercially valuable as the white sturgeon, so the fish will likely remain rare in the Lower Fraser. The one the man caught and released was probably struggling to find another mate and spawn. But Erin was excited to see that green sturgeon, to know that they were still out there, because that might mean there was still hope to understand them.

❧

Before I met Erin Stoddard and saw him catch a sturgeon, I considered fishing a profoundly dull activity. Looking at a sturgeon up close, seeing her beautiful eyes and her armored skin, I realized just how ancient this animal's anatomy truly was, and my feelings about fishing changed. "The greatest privilege of fishing," the broadcaster and author Jeremy Paxman observes, "is the obligation it puts upon you to be quietly part of a world we spend the rest of our lives trying to defy, control, or ignore." At a time when it is vitally important to listen to the world around us, fishing is another way to connect.

At a museum exhibit at the University of British Columbia's Museum of Anthropology, I recently found a map of the sturgeon's role in the culture of the Musqueam, a people who have lived on the banks of the white sturgeon's habitat for over ten thousand years. At the center of the map sat a sturgeon linked to a half dozen words, such as *territory*, *language*, and *technology*. Each word branched off into still more connections: harvesting times, place names, and intergenerational knowledge. Here was tangible evidence of all the ways that a fish touched a community and the wider culture. Dozens of diverse communities like this one were scattered up and down the coastline, and many more were around the world, each with its own unique relationship with sturgeon.

Not too long ago, Erin confirmed a new feature of white sturgeon behavior that might directly help it survive. He pinpointed the start of spawning as the third week of May until the middle of July. The angling industry agreed to halt fishing trips during that time. Here was a real-world compromise that would help sturgeon recover and preserve the fishing population for angling companies, a win that Erin could point to whenever he had to justify the time and expense of his work.

Would the average person notice the white sturgeon's extinction? "If sturgeon disappeared, it wouldn't have a huge impact on the average person," he admitted. Then he paused and reconsidered. "It would have a huge impact on the angling guides—millions of dollars are spent to go sturgeon fishing." He started to list off researchers and conservation officers who would be out of a job and all their families who would be impacted without sturgeon. He listed the Indigenous groups in the region that rely on the fish for ceremonial and cultural purposes. So, yes, those of us not directly involved might not notice the sturgeon's disappearance, but so many other loosely connected people would, and eventually all of us would feel its loss in a real way. Erin has to consider these diverse groups when he makes a decision about the fish's fate.

I first went fishing on Erin's boat as an outsider looking at a unique ecosystem of people connected through an ancient fish. As a writer, I expected to compose a well-rounded story and move on, but instead I found myself personally drawn into this world. Not long after those trips on Erin's boat, I went on a first date with a reserved German zoologist I had met online. Trying to make conversation, I brought up sturgeon. Coincidentally, the zoologist happened to work with one of the researchers on that boat, and the conversation swerved onto more familiar terrain through a shared person and an animal. A few years later, I married that zoologist. Now I count myself among those who have been forever changed by this little-known fish struggling to survive in the modern world.

Some might question the expense and effort of saving prehistoric fish. How is one ancient fish going to make a difference in the lives of a refugee, or a cruise-ship worker, or a sea turtle whose belly is full of plastic? Surely more pressing needs come before the lives of an obscure fish. What we're discovering, though, is that jettisoning any species or country over another ignores the wider world of connections. We can see it in the flags of convenience that favor individualistic gain over conserving a collective space. Our interdependence with the ocean—painful, fruitful, a part of it forever mysterious—is worth preserving. Of all the stories I witnessed, the sturgeon's connection to so many diverse people became the most hopeful sign for the ocean's, and our own, survival.

The research boat hitched to the back of the truck banged loudly over every pothole in the highway as we drove west to the ocean. Centuries from now, after the rising oceans have washed away the highway, some sturgeon might still be out in the Pacific, continuing to swim and hunt the way they have almost since they first appeared on Earth. Sturgeon have weathered at least two hundred million years, survived a few mass extinctions and a dozen ice ages. This is an infinitely more adaptable species than humans, more flexible to living and changing with the tides.

Perhaps it is we humans who are the blip. Long after we've disappeared, the water will hold the traces of the garbage we threw away, the borders we defended, the rules we broke, and the stories we told: vast and voiceless water carrying what we wanted and what we lost along the way.

EPILOGUE

THE OCEAN TODAY

Seven years into his decade-long study, Erin Stoddard continues to track sturgeon swimming through a network of thirty-five acoustic receivers in the Lower Fraser River. Unbeknownst to those fish, they have yielded important conservation wins for their kind. The study has confirmed ten spawning sites so far, and Erin's team has closed three of those sites to fishing during the reproductive and rearing months of May to July. After a placer mining operation opened on the banks of another spawning site, digging up gravel in search of gold, his research helped shut it down. Working with the Musqueam, Erin confirmed an overwintering site near a bridge that was scheduled for replacement. In the future, this discovery will hopefully help shift the construction to a time period that accommodates the nearby sturgeon.

Despite these wins, the white sturgeon population has fallen drastically between 2015, when I followed Erin in the field, and 2017, the year for which the most recent estimates are available. Ten thousand fewer white sturgeon swim in this part of the world today, a quarter fewer than the 2015 population. (There is still hope, however, that a future count will reveal inaccuracies in the sturgeon's favor.) Despite this dramatic decline, the federal government has not recognized the fish as endangered, which would help fund more research into why the fish is disappearing. Meanwhile, the sport fishery continues to grow, catching more and more sturgeon and charging hundreds of dollars a head. Huge sturgeon are caught multiple times a season. That is a lot of pressure on an struggling group of fish.

New and different threats are also emerging. Seals, for instance, have started to prey on sturgeon, driven perhaps by a lack of food in the estuary where both sturgeon and seals used to hunt eulachon. Shipping traffic is increasing in the harbor, and sturgeon suffer from propeller strikes. The number of young sturgeon between six and eight years of age is not improving. When Erin talks about managing this particularly vulnerable group of fish—the next generation who will live in the water world we're creating right now—he sounds discouraged.

As always, the bright spot for sturgeon is new and improved information. Each year Erin's study advances, more sturgeon spawn, migrate, and hunt, and Erin is watching, putting the data points together, trying to figure out what this ancient species is doing as it navigates the forces that conspire against its survival. Every year, he has new questions. He expects this trend to continue.

Technology is not the quick fix for the long and complicated troubles that ail the ocean, but when it comes to helping us learn and make decisions about sturgeon, and the whole fate of the ocean, technology is essential. There will never be enough humans to monitor the great vastness of the ocean, to watch all the ships coming and going or take endless measurements of rising acidity or shifting currents. The laws of the sea are already written, but with better and cheaper technology, we can bring about the accountability and enforcement that the ocean needs.

Recently, I came across an optimistic projection in a United Nations report about harnessing the ocean's power for collective good. If marine ecosystems are well managed, the report stated, the sea can make a massive impact on reducing poverty, fostering communities and economies, and feeding the 9 billion–some people who will live on the planet in 2050. If just one industry, like fisheries, is better managed, it can become $80 billion more productive in annual revenue each year. Taking care of the ocean means taking care of ourselves, too.

❧

Cunard provided this comment when asked about Favio Oñate Órdenes's death and the conditions on board its ships:

> We were greatly saddened by the death of Mr. Oñate Órdenes, who was a valued crew member. When the ship became aware that Mr. Oñate Órdenes was missing, the crew quickly reviewed CCTV for the open decks and found that he had jumped overboard in the early hours of the morning. The ship initiated a Mayday broadcast to which Halifax Joint Rescue Coordination Center (JRCC) responded. *Queen Mary 2* was made the On Scene Coordinator and a container ship, *M.V. Milan Express*, was requested to assist in the search. A search pattern was executed for some ten hours until, after communication with Halifax JRCC, the search was eventually halted due to fading light and reduced visibility. Very sadly, Mr. Oñate Órdenes's body could not be found. Our care team supported Mr. Oñate Órdenes's family in the period following his death.

According to Ross Klein's website, seventy-six more people have gone overboard since Favio fell off a cruise ship in August 2015.

Favio's mother, Carolina Órdenes Tobar, still hopes to mount a legal case against Cunard. Significant hurdles stand in the way, like trying to convince a lawyer in Bermuda, *Queen Mary 2*'s flag state, to mount a case against a powerful corporation. Right now the family can't afford legal costs. Favio was one of the family's main breadwinners. She continues to hope that telling his story will prevent more lives ending as his did. A few days before Christmas in 2016, a female crew member fell over the *Queen Mary 2*'s railing.

For years, Santiago kept a Facebook profile picture of him and Favio laughing on board the *Queen Mary 2*. "My memories are of the good times," he says. "I just remember him as a good friend, someone who taught me to smile in a tough situation, so that it wasn't only about work and hard times." Santiago found that even with his international work experience, it was difficult to advance in Chile's culinary world. In August 2018, he moved to Australia, where he's applying for a work permit. Part of the permit application asked for reference letters from past employers, and he asked Cunard to sign a letter outlining his role and duties on the *Queen Mary 2*. Santiago said Cunard declined to sign, because his reference letter revealed confidential information about the ship's operations. They mailed him a form recommendation letter instead.

In 2017, a trio of congressional representatives introduced legislation for the Cruise Passenger Protection Act. The bipartisan-supported legislation would bolster safety on cruise ships by requiring automatic man-overboard alert systems, allow access to video surveillance for civil lawsuits, update the Death on the High Seas Act, and address many other safety issues on board. More than two years later, the bill has not budged. If the Cruise Passenger Protection Act were passed, the millions of people who board cruise ships each year might be safer at sea, or at least better compensated when something goes wrong.

In the last few years, ocean plastic has moved from niche issue to the agenda of the 2018 G7 ministers' meeting. In the United Kingdom, a wave of zero-waste stores has opened. Up and down the coasts of North America and Africa, plastic-bag bans are being introduced. The European Union and Canada aim to ban a dozen single-use plastics, from straws to cutlery, by 2021. Accepting that the tide has turned against single-use plastics, multinational companies, like Coca Cola and Nestlé, are committing to minimum recycled content in their packaging.

"The ocean plastic awareness movement has been growing exponentially," Chloé Dubois observes, and Ocean Legacy has grown and expanded along with it. When we last spoke, Chloé was in Panama, recovering from a two-hundred-person shoreline clean she had finished leading in Mexico. She was also there to set up Ocean Legacy's first Central American recycling hub, modelled after a hub the team had opened in Vancouver in 2016. Ocean Legacy uses the Vancouver warehouse to educate school groups, run community sorting events and to process ocean plastic collected by twenty different cleanup groups. So far, the warehouse has recycled over a hundred metric tons of marine debris, most of which was heading to landfills before.

All the new attention on plastic pollution shows undeniable progress, but Chloé is still collecting the same amount of plastic in MuQwin that she always has. "Until the infrastructure is there to collect and capture these plastics, and better manage the entire planet, we're still going to find those plastics," she said. "It's a very dynamic problem. It comes down to individual action, and it comes down to much larger regulatory action—and the political will."

One challenge Ocean Legacy faces is that so much collected debris is covered in kelp or sand: a recycling nightmare, Chloé noted. It was difficult to convince companies to recycle what Ocean Legacy had recovered, so they're experimenting with setting up their own system. They purchased a plastic chipper, and Dexter, the ship mechanic, was constructing a plastic extruder that would melt down marine plastic into reusable items they could sell or use themselves. Recently, Fisheries and Oceans Canada funded Ocean Legacy's collaboration with an engineering team to recycle recovered fishing gear. Recently, they rolled out a new campaign that addresses the multiple worlds that plastic pollution affects, from policy to infrastructure development and education.

The ultimate goal is a world where Ocean Legacy wouldn't have to lead cleanups in remote jungles or build their own ocean plastic–recycling system from scratch. In the near future, Ocean Legacy is fund-raising a return to MuQwin Peninsula to lead another cleanup.

A few regulars continue to live by Ladysmith's shores, while new boats continue to come and go, staying for months or years at a time. Daniel Inkersell finally cashed in on his yacht, *Sojourn*, and sold her to a Chris Craft fan, just as he planned. He took that money and left the Patch for a trip to India. Five months after the town of Ladysmith failed to evict the Dogpatch boat-dwellers, the town succeeded in removing *Viki Lyne II* from the harbor. In May 2016, the Coast Guard towed the ship from Ladysmith and dismantled her in a nearby shipyard to the tune of $1 million. Bryan Livingstone, who still lives in the Patch, watched from his boat as the town celebrated *Viki Lyne II*'s departure. "That was really stupid," he said. "It was not going to sink for years and years, maybe a hundred and fifty years. It was a super boat. They spent a huge amount of money getting rid of that. They could have spent it on something useful, like cleaning up pollution." Bryan recently turned seventy-six, but he's not planning to leave the Patch just yet.

In early 2017, Traci Pritchard's float home, *Anchor Management*, burned down. Luckily, she was out at the time, but her two dogs died in the fire. Like many goings-on in the Patch, there were conflicting stories over whether the fire was arson or accidental. A local fund-raising effort collected just under $4,000 for Traci to restart her life. After nearly twenty years of living on water, she abandoned the water and moved back on land for good.

In January 2018, the town of Ladysmith unveiled a concept plan for its harbor. Wooden walkways, dockside restaurants and shops, bike parking, and a boutique hotel will one day line the shores of the Dogpatch if everything goes according to plan. In the illustrations, the marina will absorb the Dogpatch, creating space for two-story float homes and slips for short and long stays. Recently, the town acquired the strategic sliver of coastline that had hampered its ability to evict Dogpatch residents from the water lot. With that hurdle cleared, boat-dwellers like Bryan Livingstone might continue to argue that the town has no say over anchoring in the ocean, but that argument has not held up in court. The Dogpatch is heading into uncharted waters.

A refugee's fate moves with the tides of public opinion. Following a string of sexual assaults in Cologne and a terrorist attack at a Berlin Christmas market, German sympathy for refugees has deteriorated. A newly installed interior minister promised change and in 2018 deported nearly 25,000 asylum seekers who didn't make the government's cut. These were painful, sometimes forceful, deportations; according to a report in *Der Spiegel*, the government couldn't find enough officers to do the work. Around this time, Mohammed started a construction apprenticeship, where his boss was helping him secure a three-year work contract in Germany. One day, he came home to find an official letter informing him that his asylum request was denied and to bring his passport to a government office. Worried that he might lose his new job and apartment, he refused—and promptly lost both. According to the Dublin Regulation, Mohammed should have filed his asylum application in Italy, the first European Union country he reached. Like many migrants, he didn't, instead moving north to Germany, where he intended to settle.

A Gambian friend of Mohammed's was deported from Germany and returned to Italy to file his asylum application. Since the election of a new far-right government in Italy, the prospects for asylum-seekers there, even those with protected status, look bleak. Rescue centers are closed, homelessness is widespread, and migrants make desperate treks across the Alps. His Gambian friend ended up living on the street. "Maybe they would take me back to my country, but I would just kill myself," Mohammed says. "Being homeless in Italy is better than Ghana. Much, much better." In Germany's biggest city, he doesn't feel so constrained by his outsider status and the undocumented life he has to lead now. He works mini-jobs, a few hours here and there, passed along by a network of African immigrants. At night, he can't sleep, so he smokes cigarettes and drinks beer to numb his worries about his future, about what to do next. When I asked if he still considers himself Muslim, he said he felt torn. "I don't know which road I'm on."

In Germany's immigration system, Syrian refugees are afforded top treatment, yet life is still not easy for Hassan. In 2018, his mother was diagnosed with breast cancer. As a refugee, he has no passport, so he can't board a plane to Syria, and even if he could, it wouldn't get him to his mother's deathbed. He would be arrested for dodging military service as soon as he arrived. Nonetheless, he considered returning to see her one last time. After everything he and his family had gone through to get him to safety, he realized that ending up in a Syrian prison would truly break his parents' hearts, so he stayed in Germany and sent money home for her chemotherapy. "She was everything," he said of their close relationship. After her death, he stopped following the news from Syria—good, bad, all of it. He's lost his religion, too.

The hundreds of thousands of people who crossed into Europe on life rafts during 2015 and 2016 have led the world to the political crossroads of today. Across Europe, a wave of populist anti-immigrant groups rode into power in Poland, Italy, and Hungary. Others have gained prominence in governments across the continent. Similar anti-immigration anxiety sparked the disastrous Brexit proceedings in the United Kingdom. The long tenure of German chancellor Angela Merkel, who allowed nearly one million asylum-seekers to enter Germany, is coming to an end. At the US–Mexican border, a bottleneck of asylum seekers from Central America grows longer, President Trump continues to push for a border wall, families are separated, and six children have died in detention as of this writing. Europe's refugee crisis was not an isolated incident after all. As more and more people are displaced by climate change, extreme poverty, violence, and disaster, the debate over who is entitled to refuge will continue to rage.

Sailing offshore is the objective of all the Bluewater training, yet so few sailors I've met linger on describing the crossing. According to Fiona McGlynn, that's because most sailors find it physically uncomfortable, herself included. During the twenty-six-day passage to the Marquesas, she remembers, her body felt tense the whole time. Even sitting upright in a chair as the boat leaned to one side caused an omnipresent strain. There was also the mental strain of feeling constantly alert and on edge at sea. Eventually, twenty days in, she relaxed enough to read a book. Her brain felt empty and clear from so many days without computer screens and the stimulating pulse of the internet.

After reaching the Marquesas, Fiona and Robin spent the next eight months island-hopping across the South Pacific. The usual advice is to spend two years sailing through this stretch of ocean, but like many younger sailors pinched for cash, they crossed quickly. That was an intense undertaking, Fiona recalled. They'd cruise into a picturesque atoll, recover from the journey, make any boat repairs, and ship off again before the next weather window closed. Along the way, they turned the experience into a source of income, selling stories to sailing magazines and running a blog.

By the time they reached Australia, they were dreaming of life on land. Looking at the map of the Pacific, it's astounding to think how much water they'd sailed through. "You don't realize how big the ocean is until you try to cross it in a sailboat," said Fiona. They sold the *MonArk* in Australia. In early 2018, they returned to Canada, settling in a cabin near the Alaska–British Columbia border. They started a garden, grew their own mushrooms, and adopted two rescue huskies. They would change almost nothing about that trip, she said—but they would have sailed a bigger boat.

❧

Since my visit to Pete Romano's warehouse in Los Angeles, he has continued at his same pace, shooting commercials, television, and four to five big-budget movie productions each year. Some recent top-shelf names he's worked with include Quentin Tarantino ("He's good with the crew"); Brad Pitt ("probably shot the spookiest underwater sequence I've ever done"); and Ang Lee ("I'd do anything anywhere with him"). But the art of underwater cinematography in southern California took a hit when Tank One Studios closed. The owner, who suffered from health problems, moved away and eventually shut down the business altogether.

The 2018 box office hit *Aquaman* relied on CGI for most of the underwater scenes. Actors mounted on cables were flown through a stage area, while watery flourishes and sea turtles were added later in postproduction. The result is an underwater movie that looks strangely like it's happening in outer space. Characters hover, rather than float, and glide instead of swim. Still, Pete is no longer quite so adamant that CGI will never capture the look and feel of a person swimming underwater. "Maybe the average Joe doesn't really care, but I still think it somehow has a cartoonish feel to it when they do the whole thing CGI," Pete said. "We have a little way to go. I might have to shoot a few more years before it all implodes."

~

Over the last few years, a lot has happened in the life of the sea. In 2010, the United Nations set a goal of placing ten percent of the world's ocean in marine protected areas by 2020; that must have seemed a world away at the time. As we inch closer to the end of the decade and consider our scorecard, we've only protected less than five percent of the ocean in any meaningful way, according to a study published in the journal *Marine Policy*. That has not stopped governments from claiming success, even when destructive fishing continues in so-called protected areas. The deadline has since shifted; the International Union for Conservation of Nature has asked its government members to protect 30 percent of the ocean by 2030. Experts predict that by that time, plastic pollution in the ocean will have doubled; two thirds of the fish we eat will be farmed, much of it at sea; offshore wind will lead power-generation technology; and an international consortium of oceanographers will have mapped the entire seafloor. Skip ahead twenty years to 2050, and nearly every seabird is predicted to have plastic in its gut. Seaborne trade, at twelve billion metric tons in 2018, is expected to quadruple. Climate change could displace around 200 million people. By 2100, the future gets a little more hazy. Depending on how quickly ice sheets in Antarctica and Greenland melt, the oceans will likely rise two feet (one meter) by 2100, if not more, sinking cities from Bangkok to Boston. By that year, 50 percent of the sea, thanks to climate change's impact on phytoplankton, will be a more vibrant emerald or a deeper blue.

Reading through these projections, it's easy to feel pessimistic about the future, about how much we've manhandled the ocean and hurt ourselves in the process. But I take heart knowing that these timelines even exist. A decade earlier, we didn't take the ocean's temperature quite so often and in so many ways. Today there's hunger for insight about what's ahead. If the status quo continues, where is the world headed? Who will fare the best—and the worst? Of course, most of these numbers are a guessing game based on business-as-usual scenarios. They could never account for all the unknown efforts around the world that will lift up or drag down our progress along the way.

Scientists and environmentalists often use the canary-in-the-coal-mine analogy when they highlight the importance of an endangered animal. Take sharks, for instance. They are some of the most uniquely threatened animals in the ocean, but why should you personally care about sharks? Because the health of a top predator, such as a shark, indicates the health of the ecosystem beneath it and alerts us to threats that might impact us. The shark is the canary in the coal mine, an early-warning system that gives us a chance to respond before it's too late.

The ocean is a multitude, linking us to one another and to every other living organism on the planet in unimaginable ways. Every day we write a story about the ocean that will have an impact on lives far beyond our own. For our survival and the ocean's, it is time to imagine all the ways in which our stories are linked. In the end, the ocean's story is also our own.

ACKNOWLEDGMENTS

In the four years I spent writing and reporting *The Imperiled Ocean*, I spoke to dozens of people who gave their time, knowledge, and insight to this project.

Thank you to Pete Romano and his staff for letting me explore the Hydroflex warehouse in Los Angeles. Thanks also to the now-shuttered Tank One Studios in Long Beach, including owner Al Gerbino, handyman and safety diver Joseph Smith, producer Michael Park, and administration assistant Melody Cortez. Water stunt coordinator Katie Rowe talked me through the challenges of working with water on set. Animators Eloi Andaluz and Will Wallace provided insight into the evolving industry of visual effects.

The Bluewater Cruising Association was a phenomenal resource while I wrote about the world of offshore sailing. Thank you to Bluewater sailors Dennis Giraud, Gillian West, Helen and Jean Baillargeon, Beth and Norm Cooper, and Fiona McGlynn and Robin Urquhart for letting me board their boats and understand their lives.

Hassan Basheer and Mohammed Botwe are pseudonyms for the two men whose Mediterranean journeys I covered. Both relived the most difficult moments in their lives by retelling their stories, and I thank them for that sacrifice. Neither wanted to be identified out of fear of retaliation against them or their families or of legal consequences in Germany. I interviewed a half dozen other refugees from Afghanistan, Syria, Guinea, and Somalia. For similar reasons I won't identify them, but their stories and sacrifices helped bring this story to the page. The German nonprofit Pestalozzi Kinder- und Jugenddorf helped connect me with recently arrived refugees. Social worker Karen Hansen, who worked at the refugee house where Mohammed lived temporarily, answered my questions, as did Elisa Kaltenbach at the University of Constance's Department of Psychology.

Thanks go to the residents of Ladysmith and the Dogpatch, including Traci Pritchard, Bryan Livingstone, Ken Ireland, Daniel Inkersell, Lew McArel, Vince Huard, Luke English, Craig Appleyard, and Brian Charron. At the Ladysmith Maritime Society docks, thank you to Brenda and Norman Brook and to Rod Smith, the wharfinger. Ladysmith's mayor, Aaron Stone, and city manager, Ruth Malli, also answered my questions, as did RCMP officer Ken Brissard.

Ocean Legacy's Chloé Dubois, James Middleton, and Eric (Dexter) McGillivray included me on their 2015 expedition to MuQwin Peninsula. Along the way, Julian (Garuda) Noel, Eric Robinson, and many others associated with the nonprofit's efforts contributed to writing and reporting. Thank you to oceanographer Kara Lavender Law and marine biologist Chelsea Rochman for their expert input.

Cunard declined to be interviewed. In lieu of official information, Favio Oñate Órdenes's friends and family helped fill out his life and time aboard, including Santiago Gutiérrez, Carolina Órdenes Tobar, and Carolina Oñate Órdenes. Author and former cruise-ship waiter

Brian David Bruns related his personal experiences working on all of the Big Three cruise ships and helped confirm information about life on board. Thanks also to sociologist Ross Klein and lawyer James Walker for their insights.

Thank you to Erin Stoddard for allowing me on board his research boat and answering my endless questions about the life and future of sturgeon. Garrett Martingale, Jeff Beardsdale, Owen Catherall, Steve McAdams, and Greg George were also on board and provided diverse perspectives on studying and conserving a mysterious fish. Thanks also to Fred Whoriskey, executive director of the Ocean Tracking Network at Dalhousie University, Sarah Schreier at the Fraser River Sturgeon Conservation Society (FRSCS), and Marvin Rosenau, instructor in the Fish, Wildlife and Recreation program at the British Columbia Institute of Technology.

Thank you to the many colleagues and friends who read this work in progress and contributed invaluable feedback, including my thesis supervisor, Timothy Taylor, and thesis committee members Kevin Chong and Deborah Campbell at the University of British Columbia's Creative Writing Program. I owe a huge debt of gratitude to Andreas Schroeder for giving me the space and time to write in his cabin. Jeremy Heywood, Catherine Po, Jane Campbell, Sierra Skye Gemma, Ryan Huizing, Alannah Biega, Chlöe Ellingson, Till Harter and Mark Gilligan also contributed to this project in ways big and small, as did my parents, Carole Prest and Paul Trethewey.

Thanks go to the editorial teams at the *Walrus* and *Hakai Magazine*, which published versions of chapters 4 and 5, and to my editor Jill Ainsley, whose handiwork improved every line in this book. Many more unnamed people contributed along the way through short conversations or through interviews that never made it to the page. But their help was essential in piecing together an overview of the ocean that is much, much bigger than one individual narrative. I am humbled by their generosity and thank each one for believing in me and this project.

Thank you, Till Harter, for your endless intelligence and awareness about the ocean world. On our first date we talked about sturgeon, and you've made this book far smarter than it ever would have been without you.

SOURCES

INTRODUCTION

The websites of the National Ocean Service, the United Nations Environment Programme, the Food and Agriculture Organization of the United Nations, NASA Science, and the Pacific Marine Environmental Library provided valuable facts and figures. On National Geographic's website, you can watch oceanographer Bob Ballard explain the discovery and significance of hydrothermal vents. The websites CruiseLawNews.com, ParisMoU.org, and SeaShepherdGlobal.org were also helpful.

I learned about early man's seafood diet from Jon Erlandson and Scott Fitzpatrick's article "Oceans, Islands, and Coasts" (*Journal of Island and Coastal Archaeology*, 2006) and Curtis Marean's "Making the paper" (*Nature*, 2007). Estimates about plastic in the ocean come from Jenna R. Jambeck's "Plastic waste inputs from land into the ocean" (*Science*, 2015) and the Ellen MacArthur Foundation's "The New Plastics Economy" (2016), among other studies and reports.

Books I consulted include Lincoln Paine's *The Sea and Civilization* (2015); Alanna Mitchell's *Sea Sick* (2010); William Langewische's *The Outlaw Sea* (2004); Wallace J. Nichol's *Blue Mind* (2015); and Rose George's *Ninety Percent of Everything* (2014). Garrett Hardin's groundbreaking essay "The Tragedy of the Commons" was published in *Science* in 1968. I also drew on reporting by the *Guardian*, *Antigua Observer*, the *New York Times*, *Newsweek*, *Forbes*, the BBC, and MIT News. The quote from Anne Stevenson is from her poem "The North Sea off Carnoustie."

ONE: CAPTURING THE WATER WORLD

On YouTube, you can watch clips from vintage underwater films and TV shows, including *Sea Hunt*, Jacques Cousteau's groundbreaking TV series *The Undersea World of Jacques Cousteau*, *Creature from the Black Lagoon*, and *Adventures in the Red Sea*. YouTube was also my purveyor of water-based music videos and tutorials on animating water. The detailed, behind-the-scenes breakdown of animated water effects on film websites like ArtoftheTitle.com, FXGuide.com, and AWN.com also helped.

Books I consulted include Chet Van Duzer's *Sea Monsters on Medieval and Renaissance Maps* (2013); Joseph Nigg's *Sea Monsters* (2013); James Nestor's *Deep* (2014); Sebastian Junger's *The Perfect Storm* (1997); and Marcus Rediker's *Between the Devil and the Deep Blue Sea* (1987). J. Roberta Coffeit's PhD dissertation, "She 'Too much of water hast': Drownings and Near-Drownings in Twentieth-Century American Literature by Women" (University of North Texas, 2001), was also helpful. Reporting in *Smithsonian Magazine*, *Hakai Magazine*, the *Guardian*, and the *New Yorker* provided background information on diverse related topics.

TWO: SAILING THE SOUTH PACIFIC

Many of the sailors I interviewed are also dedicated writers, and I used their accounts to reconstruct experiences at sea, including Beth and Norm Cooper's blog, *Voyages of Sarah Jean II*, and Beth's articles in *Blue Water Sailing*; Fiona McGlynn and Robin Urquhart's blog, Waterborne, and Fiona's writing in *Canadian Geographic*, *Pacific Yachting*, and other publications. Sailing magazines and associations provided up-to-date information on offshore sailing, including *Pacific Yachting*, *Yachting World*, the American Sailing Association, Cruising World, Blue Water Sailing, Sail Feed, and too many others to list.

Books that inspired or informed this chapter include Joshua Slocum's *Sailing Alone Around the World* (1900); Bernard Moitessier's *The Long Way* (1971); Don Holm's *The Circumnavigators* (1974); Derek Lundy's *The Godforsaken Sea* (2000); Ron Hall and Nicholas Tomalin's *The Strange Last Voyage of Donald Crowhurst* (1970); Joseph Conrad's *Mirror of the Sea* (1906); Cheryl Strayed's *Wild* (2013); Hal Roth's *How to Sail Around the World* (2003); *Jimmy Cornell's World Cruising Routes* (2014, 7th ed.); and last but certainly not least, Kevin Patterson's *The Water in Between* (1999).

The websites of the National Oceanic and Atmospheric Administration, the *Canadian Encyclopedia*, and MarineTraffic.com came in good use, as did reporting in *Hakai Magazine*. Halifax historian Joel Zemel helped clarify my great-great-grandfather's role in the events leading up to the Halifax Explosion. The 2013 film *All Is Lost* and the Anarchist Yacht Club's 2007 documentary *Hold Fast* portrayed the nuances in cruising. I learned about the spectrum of failure from Malcolm Gladwell's essay "The Art of Failure" (*New Yorker*, 2000), which led me to the 2017 study "Curtailing Cascading Failure" in *Science*. The quote from Annie Dillard comes from her book *Pilgrim at Tinker Creek* (1974).

THREE: CROSSING THE MEDITERRANEAN

Patrick Kingsley's reporting in the *Guardian* in 2015 and 2016 was essential for the in-the-field details about smuggling fees and boats, as was his book *The New Odyssey* (2017). I also drew on reporting in the *New Yorker*, *Time*, CNN, BBC, CBS, Deutsche Welle, the

Jerusalem Post, the *Washington Post*, the *New York Times*, the *Economist*, the *Daily Telegraph*, and the *Guardian*. "The Migrants' Files," a Journalism++ collection of reporting across fifteen countries throughout 2014 and 2015, was also a useful hub of information.

The website and resources of WatchtheMed.net, Frontex, the United Nations High Commissioner for Refugees, and the International Organization for Migration provided facts, figures, and statistics. Books I consulted include Charlotte McDonald-Gibson's *Cast Away* (2016); the anthologies *Sea Changes*, edited by Bernhard Klein and Gesa Mackenthun (2004), and *Migration by Boat*, edited by Lynda Mannik (2016); and Lincoln Paine's *The Sea and Civilization* (2015). The quote from Homer's *The Odyssey* comes from Emily Wilson's 2017 translation.

FOUR: FLOATING FREE

I owe a debt to local news sources, including the *Ladysmith Chemainus Chronicle*, *Squamish Chief*, *Take 5*, CHEK, and *Oakland North*. I also relied on reporting by the CBC, the *New York Times*, the *Guardian*, *Hakai Magazine*, and the *Financial Times*. The Canada Shipping Act outlines the rules for disposing and salvaging boat wrecks. Duhaime's *Encyclopedia of Law* helped with legal terminology, as did Darren Williams's article on unwanted vessels, "Anchorage vs. Moorage" (www.duhaime.org, 2010). I drew real estate information from TheLandingSF.com, Booking.com, Amsterdamhouseboats.nl, and ThamesRiverHomes.co.uk. The websites of the United Nations, the Canadian Parliament, the Government of Canada, the Government of British Columbia, Transport Canada, the Migration Policy Institute, Islands Trust, Town of Ladysmith, Billion Oyster Project, and Georgia Strait Alliance also supplied relevant information.

Books referenced include Penelope Fitzgerald's novel *Offshore* (1971) and Wallace J. Nichols's *Blue Mind* (2015). I found out about the impact of water quality on real estate prices from Janne Artell's study, "Water Quality Has a Significant Impact on Recreational

Value and Waterfront Prices" (MIT Agrifood Research Finland, 2013). The E. P. Thompson quote is from *The Making of the English Working Class* (1964).

FIVE: CLEANING THE COAST

National Geographic, CTV News, Global News, *Smithsonian Magazine*, the *Independent*, *National Post*, the *L.A. Times*, the *New York Times*, *Hakai Magazine*, *Tree Hugger*, *Front Line*, and www.howstuffworks .com were useful sources. I drew additional information from the websites of Environment Canada, the American Chemistry Council, the National Oceanic and Atmospheric Administration and its Marine Debris Group, the Food and Agriculture Organization of the United Nations, Beachcombers Alert, *Deep Sea News*, and the World Society for the Protection of Animals.

Susan Freinkel's *Plastic* (2011) and Sheila Watt-Cloutier's *The Right to Be Cold* (2015) were also helpful. I drew on the published work of a number of scholars and research teams studying plastic pollution and marine debris, including David K. A. Barnes, Jonathan V. Durgadoo, Alexander Ter Halle, Jenna Jambeck, Bum Gun Kwon, Joleah B. Lamb, Kara Lavender Law, Jongsu Lee, Chelsea Rochman, and Madeleine Smith. I am also indebted to the work of oceanographer Curtis Ebbesmeyer and sustainability professor Jem Bendell.

SIX: CRUISING THE NORTH ATLANTIC

The websites CruiseJunkie.com and CruiseLawNews.com were invaluable. Jon Ronson's essay "Rebecca Coriam: Lost at Sea," published in the *Guardian* in 2011, was inspirational. I also drew on reporting in the *Daily Telegraph*, the *Chronicle Herald*, the *New York Times*, and by NPR. The 2016 series "Vacations in No Man's Sea" by Univision and the Columbia Journalism School is the most valuable and extensive reporting to date on a range of cruise-ship activities at sea and on land. I also consulted with two Univision reporters, Maya Primera and Almudena Toral, about what they uncovered.

Although I did not have access to the *Queen Mary 2*, I did tour the *Queen Mary* at her permanent home in Long Beach and "toured" the *Queen Mary 2* using personal videos shot by passengers and staff and posted on YouTube. Carolina Tobar provided copies of her son's death certificate, the Government of Bermuda's investigation, and a report from the Chilean Regional Office of the Prosecutor on its search of *Queen Mary 2* while it was in port in Valparaiso, Chile. Carolina Oñate Órdenes provided copies of a letter of complaint that she and her family sent to Cunard.

p. 136: "In the magazine *California Sunday*, ProPublica reporter Lizzie Presser wrote of a Filipino waiter on a Carnival cruise ship who worked as much as fourteen hours some days . . ." Lizzie Presser, "Below Deck," *California Sunday*, February 2, 2017, https://story.californiasunday.com/below-deck.

p. 144: ". . . *Queen Mary 2* has one of the more generous staff-to-passenger ratios in the industry, with 1,292 employees serving 2,691 passengers, according to Cunard's website." https://www.cunard.com/en-us/cruise-ships/queen-mary-2/9.

p. 145: "The majority of cruise-ship workers are Filipino, according to researcher William C. Terry . . ." William C. Terry "Working on the Water: On Legal Space and Seafarer Protection in the Cruise Industry," Economic Geography, 85:4, 463-482.

p. 145: ". . . while other low-ranking jobs are filled by workers from India or Indonesia, according to a paper published by Bin Wu at Cardiff University's Seafarers International Research Center." Bin Wu, "The World Cruise Industry: A Profile of the Global Labour Market," Seafarers International Research Center, Cardiff University, 2005.

p. 145: "In 2016, an investigative reporter from Univision Noticias, an American-Spanish language broadcast network, described going below deck on a Carnival ship to see where staff . . ." Damià S. Bonmatí, "Sweatshop on the High Seas" Univision Noticias and Columbia Journalism School, June 26, 2016, http://huelladigital.univision noticias.com/cruceros-vacaciones-en-aguas-de-nadie /trabajo/index-lang=en.html.

p. 145: "Newspapers like *The Guardian* and academics like Bin Wu and Christine Chin . . ." Gywn Topham, "P&O cruise ship staff paid basic salary of 75p an hour," *The Guardian*, April 29 2012, https://www.theguardian .com/business/2012/apr/29/cruise-firm-performance -bonuses-tips.

Bin Wu, "The World Cruise Industry: A Profile of the Global Labour Market," Seafarers International Research Center, Cardiff University, 2005.

Christine B.N. Chin, "Labour Flexibilization at Sea: 'MINI U[NITED] N[ATIONS]' CREW ON CRUISE SHIPS," International Feminist Journal of Politics, 2008. https://www.tandfonline.com/doi/full /10.1080/14616740701747584.

p. 145: "ProPublica reporter Lizzie Presser traced the hiring practices of cruise ships to the colonial ties between countries in *California Sunday* . . ." Lizzie Presser, "Below Deck," *California Sunday*, February 2, 2017, https://story .californiasunday.com/below-deck.

p. 145: "In a 2019 story, *Business Insider* interviewed 35 current and former workers and found that pay can range from $500 to $10,000 . . ." Mark Matousek, "Cruise-ship workers reveal how much money they really make," *Business Insider*, June 3, 2019, https://www.businessinsider.com/cruise -ship-workers-reveal-how-much-money-they-make-2019-5.

p. 147: "A Cunard press release justified the move as a step into the wedding-at-sea market that the Union Jack forbid because the United Kingdom restricts legal weddings to churches, registry offices, and other approved venues, as Cruise Critic and other cruise coverage sites reported."

"It's Official: Cunard Re-flags Cruise Ships in Bermuda, Launches Weddings at Sea," Cruise Critic, October 19, 2011, https://www.cruisecritic.com/news/news.cfm?ID=4634.

"Cunard confirms change of flag," Captain Greybeard, October 19, 2011, https://www.captaingreybeard.com/2011/10/cunard-confirms-change-of-flag.html.

p. 148: "These countries ask few questions, impose little regulation, and offer generous free-market latitude, an issue that the International Transport Workers' Federation and other academics have covered for years.

"Flags of Convenience" International Transport Workers' Federation, https://www.itfglobal.org/en/sector/seafarers/flags-of-convenience.

Elizabeth R. DeSombre, *Flagging Standards: Globalization and Environmental, Safety, and Labor Regulations at Sea* (Cambridge: MIT Press, 2006), 3.

p. 150: ". . . an English childcare supervisor abused over a dozen boys on all three of the Cunard ships where he worked, according to BBC reporting."

"Cunard cruise ship worker Paul Trotter admits abusing children," *BBC*, April 20, 2012, https://www.bbc.com/news/uk-england-wiltshire-17784458.

p. 151: "Every year, between fifteen and twenty-five people fall off a cruise ship, according to sociologist Ross Klein's ongoing tally on his website CruiseJunkie.com."

"Events at Sea: All the Things That Can Wrong on a Cruise," Cruise Junkie, http://www.CruiseJunkie.com/events.html.

SEVEN: SAVING A PREHISTORIC FISH

The websites of Fisheries and Oceans Canada, the Committee on the Status of Endangered Wildlife in Canada, the British Columbia Ministry of Water, Land and Air (as it was then called), the International Union for Conservation of Nature, the World Wildlife Federation, the Smithsonian, the Fraser River Conservation Society, the Upper Columbia White Sturgeon Recovery Initiative, the University of California San Diego, and Global Fishing Watch provided essential information.

Reporting by *Vancouver Magazine*, the *Oregonian*, the *New York Times*, Reuters, *Tampa Bay Times*, the *Seattle Times*, *Hakai Magazine*, *Nautilus*, *Vancouver Sun*, *Eugene Weekly*, and Jefferson Public Radio was also useful. In 2004 the Fraser River Conservation Society held a forum on the white sturgeon that brought together dozens of sturgeon stakeholders from government, Indigenous groups, fisheries, and interested citizens. I relied on the published report from that forum.

I also found the work of Ricardo Betancur-R., Evelyn R. Labour, Montana F. McLean, Antoine Richards, K. J. Sulak, and Chris S. M. Turney (and the researchers who collaborate with them) incredibly helpful. Matthew D. Evenden's *Fish versus Power* (2004) provided useful context. The Musqueam's sturgeon-knowledge maps were part of the 2018 exhibit *Culture at the Center* at the Museum of Anthropology at the University of British Columbia. The Jeremy Paxman quote comes from his book *Fish, Fishing and the Meaning of Life* (1994).